Edited by
Jean-Christophe Hadorn

**Solar and Heat Pump Systems
for Residential Buildings**

Related Titles

Athienitis, A., O'Brien, W. (eds.)

Modeling, Design, and Optimization of Net-Zero Energy Buildings

2015

Print ISBN: 978-3-433-03083-7

Hens, H.S.

Performance Based Building Design 1

From Below Grade Construction to Cavity Walls

2012

Print ISBN: 978-3-433-03022-6

Hens, H.S.

Performance Based Building Design 2

From Timber-framed Construction to Partition Walls

2013

Print ISBN: 978-3-433-03023-3

Eicker, U.

Energy Efficient Buildings with Solar and Geothermal Resources

2014

Print ISBN: 978-1-118-35224-3

Edited by
Jean-Christophe Hadorn

Solar and Heat Pump Systems for Residential Buildings

Editor

Jean-Christophe Hadorn
BASE Consultants SA
8 rue du Nant
1211 Geneve 6
Switzerland

Cover: The cover photo shows selective unglazed collectors in Vouvry (CH). (Photo by O. Graf, © Energie Solaire SA Switzerland)

■ All books published by **Ernst & Sohn** are carefully produced. Nevertheless, authors, editors, and publisher do not warrant the information contained in these books, including this book, to be free of errors. Readers are advised to keep in mind that statements, data, illustrations, procedural details or other items may inadvertently be inaccurate.

Library of Congress Card No.: applied for

British Library Cataloguing-in-Publication Data
A catalogue record for this book is available from the British Library.

Bibliographic information published by the Deutsche Nationalbibliothek
The Deutsche Nationalbibliothek lists this publication in the Deutsche Nationalbibliografie; detailed bibliographic data are available on the Internet at <http://dnb.d-nb.de>.

© 2015 Wilhelm Ernst & Sohn, Verlag für Architektur und technische Wissenschaften GmbH & Co. KG, Rotherstraße 21, 10245 Berlin, Germany

All rights reserved (including those of translation into other languages). No part of this book may be reproduced in any form – by photoprinting, microfilm, or any other means – nor transmitted or translated into a machine language without written permission from the publishers. Registered names, trademarks, etc. used in this book, even when not specifically marked as such, are not to be considered unprotected by law.

Print ISBN: 978-3-433-03040-0
ePDF ISBN: 978-3-433-60484-7
ePub ISBN: 978-3-433-60485-4
Mobi ISBN: 978-3-433-60482-3
oBook ISBN: 978-3-433-60483-0

Typesetting: Thomson Digital, Noida, India
Printing and Binding: betz-druck GmbH, Darmstadt, Germany

Printed in the Federal Republic of Germany
Printed on acid-free paper

Contents

	About the editor and the supervisors	IX
	List of contributors	XI
	IEA solar heating and cooling programme	XV
	Forewords	XVII
	Acknowledgments	XIX
1	**Introduction**	1
1.1	The scope	1
1.2	Who should read this book?	1
1.3	Why this book?	1
1.4	What you will learn reading this book?	2
	Internet sources	4

Part One
Theoretical Considerations 5

2	**System description, categorization, and comparison**	7
2.1	System analysis and categorization	7
2.1.1	Approaches and principles	7
2.1.2	Graphical representation of solar and heat pump systems	8
2.1.3	Categorization	9
2.2	Statistical analysis of market-available solar thermal and heat pump systems	11
2.2.1	Methods	12
2.2.2	Results	14
2.2.2.1	Surveyed companies	14
2.2.2.2	System functions	14
2.2.2.3	System concepts	15
2.2.2.4	Heat pump characteristics – heat sources	15
2.2.2.5	Collector types	17
2.2.2.6	Cross analysis between collector type and system concept	18
2.3	Conclusions and outlook	19
2.4	Relevance and market penetration – illustrated with the example of Germany	19
	References	21
3	**Components and thermodynamic aspects**	23
3.1	Solar collectors	23
3.2	Heat pumps	28
3.3	Ground heat exchangers	34
3.3.1	Modeling of vertical ground heat exchangers	38
3.3.2	Modeling of horizontal ground heat exchangers	40
3.3.3	Combining GHX with solar collectors	41
3.4	Storage	42

3.4.1	Sensible heat storage and storage in general	42
3.4.2	Latent storage	45
3.4.3	Thermochemical reactions and sorption storage	46
3.5	Special aspects of combined solar and heat pump systems	47
3.5.1	Parallel versus series collector heat use	47
3.5.2	Exergetic efficiency and storage stratification	50
	References	52

4 Performance and its assessment — 63

4.1	Introduction	63
4.2	Definition of performance figures	65
4.2.1	Overview of performance figures in current normative documents	65
4.2.1.1	Heat pumps	66
4.2.1.2	Solar thermal collectors	67
4.2.2	Solar and heat pump systems	67
4.2.3	Efficiency and performance figures	68
4.2.4	Component performance figures	70
4.2.4.1	Coefficient of performance	70
4.2.4.2	Seasonal coefficient of performance	70
4.2.4.3	Solar collector efficiency	71
4.2.5	System performance figures	71
4.2.5.1	Seasonal performance factor	71
4.2.6	Other performance figures	72
4.2.6.1	Solar fraction	72
4.2.6.2	Renewable heat fraction	74
4.2.6.3	Fractional energy savings	74
4.3	Reference system and system boundaries	75
4.3.1	Reference SHP system	75
4.3.2	Definition of system boundaries and corresponding seasonal performance factors	77
4.4	Environmental evaluation of SHP systems	87
4.4.1.1	Primary energy ratio	90
4.4.1.2	Equivalent warming impact	91
4.4.1.3	Fractional primary energy savings	91
4.4.1.4	Fractional CO_2 emission savings	91
4.5	Calculation example	91
	Appendix 4.A Reviewed standards and other normative documents	97
	References	102

5 Laboratory test procedures for solar and heat pump systems — 103

5.1	Introduction	104
5.2	Component testing and whole system testing	106
5.2.1	Testing boundary and implications on the test procedures	106
5.2.2	Direct comparison of CTSS and WST	109

5.2.3	Applicability to SHP systems	112
5.2.4	Test sequences and determination of annual performance	115
5.2.4.1	Direct extrapolation of results (WST for combi-systems)	116
5.2.4.2	Modeling and simulation	117
5.2.5	Output	119
5.3	Experience from laboratory testing	120
5.3.1	Extension of CTSS test procedure toward solar and heat pump systems	120
5.3.2	Results of whole system testing of solar and heat pump systems	121
5.3.2.1	Excessive charging of the DHW zone	123
5.3.2.2	Exergetic losses in general	123
5.3.3	Extension of DST test procedure toward solar and heat pump systems	123
5.4	Summary and findings	126
	References	128

Part Two
Practical Considerations 131

6	**Monitoring**	133
6.1	Background	133
6.2	Monitoring technique	134
6.2.1	Monitoring approach	134
6.2.2	Measurement technology	137
6.2.2.1	Data logging systems	137
6.2.2.2	Heat meters	137
6.2.2.3	Electricity meters	138
6.2.2.4	Meteorological data	139
6.2.2.5	Temperature sensors	139
6.3	Solar and heat pump performance – results from field tests	139
6.4	Best practice examples	145
6.4.1	Blumberg	145
6.4.2	Jona	147
6.4.3	Dreieich	149
6.4.4	Savièse	152
6.4.5	Satigny	154
	References	157

7	**System simulations**	159
7.1	Parallel solar and heat pump systems	159
7.1.1	Best practice for parallel solar and heat pump system concepts	161
7.1.2	Performance of parallel solar and heat pump systems	164
7.1.3	Performance in different climates and heat loads	166
7.1.4	Fractional energy savings and performance estimation with the FSC method	169
7.2	Series and dual-source concepts	171

7.2.1	Potential for parallel/series concepts with dual-source heat pump	171
7.2.2	Concepts with Ground Regeneration	173
7.2.3	Other series concepts: dual or single source	177
7.2.4	Multifunctional concepts that include cooling	183
7.3	Special collector designs in series systems	184
7.3.1	Direct expansion collectors	184
7.3.2	Photovoltaic–thermal collectors	184
7.3.3	Collector designs for using solar heat as well as ambient air	186
7.4	Solar thermal savings versus photovoltaic electricity production	187
7.5	Comparison of simulation results with similar boundary conditions	188
7.5.1	Results for Strasbourg SFH45	189
7.5.1.1	Heat sources	191
7.5.1.2	System classes	191
7.5.1.3	Dependence on collector size and additional effort	192
7.5.1.4	Electricity consumption	193
7.5.2	Results for Strasbourg SFH15 and SFH100	195
7.5.3	Results for Davos SFH45	196
7.6	Conclusions	197
	Appendix 7.A Appendix on simulation boundary conditions and platform independence	199
	References	204
8	**Economic and market issues**	**209**
8.1	Introduction	209
8.2	Advantages of SHP systems	210
8.3	The economic calculation framework	211
8.4	A nomograph for economic analysis purposes	216
8.5	Application to real case studies	219
	References	228
9	**Conclusion and outlook**	**229**
9.1	Introduction	229
9.2	Components, systems, performance figures, and laboratory testing	229
9.3	Monitoring and simulation results and nontechnical aspects	231
9.4	Outlook	233
9.4.1	Energy storage	233
9.4.2	System prefabrication	234
9.4.3	System quality testing	234
9.4.4	Further component development	234
	Glossary	237
	Index	241

About the editor and the supervisors

Editor

Jean-Christophe Hadorn started his career as a researcher on large-scale storage of solar heat in deep aquifers (1979–1981). Since several years, Mr. Hadorn has been appointed as External Manager of Thermal Solar Energy and Heat Storage Research Program by the Swiss government. Mr. Hadorn was a participant in IEA SHC Task 7 on "Central Solar Heating Plants with Seasonal Storage" (1981–1985) and initiated Task 26 on "Solar Combisystems" (1996–2000). He was Operating Agent of Task 32 on "Heat Storage" (2003–2007). Since 2000, he leads an engineering company and designs solar thermal and PV plants. In 2010, he was chosen by an international committee of the IEA as the Operating Agent for the IEA SHC Task 44 "Solar and Heat Pump Systems" also supported by the Heat Pump Programme under Annex 38, project that produced this book.

Supervisors

Dr. Matteo D'Antoni is a senior researcher working at the Institute for Renewable Energy of the European Academy of Bolzano (EURAC) in Italy. He is active in the development of hybrid renewable energy systems for residential and commercial applications and in the design of building integrated solar thermal technologies. He is an expert in numerical calculus and transient simulation of energy systems. He managed EU funded and industry commissioned projects. Dr. D'Antoni lectures designers on the topic of renewable energy sources and energy system simulations.

Dr. Michel Y. Haller is Head of Research at the Institut für Solartechnik SPF at the University of Applied Sciences HSR in Switzerland. He holds a title of Master in Environmental Sciences of ETH Zürich, and obtained his doctoral degree in engineering at Graz University of Technology. He is coordinator of the EU project MacSheep, and author of more than 50 refereed papers. Dr. Haller was leader of the Subtask C "Modelling and Simulation" of the IEA SHC Task 44/HPP Annex 38 on "Solar and Heat Pump Systems."

Sebastian Herkel is a senior researcher at the Fraunhofer Institute for Solar Energy Systems ISE in Freiburg, Germany and Head of Solar Buildings Department. He is a researcher working in applied research in building energy performance and renewable energy systems. His focus is on integral energy concepts for buildings and neighborhoods, scientific analysis of building performance, and technologies for integration of renewable energy in buildings.

Ivan Malenković is a researcher at the Fraunhofer Institute for Solar Energy Systems ISE with over 10 years of experience in R&D, testing, and standardization in the field of heat pumping technologies. He is currently responsible for the ServiceLab for performance evaluation within the Competence Centre for Heat Pumps and Chillers at ISE. He participated in a number of IEA SHC and HPP Tasks and Annexes and was a subtask leader in IEA SHC Task 44/HPP Annex 38. He is author and co-author of a number of articles published in conference proceedings and reviewed journals.

Christian Schmidt is a researcher at the TestLab Solar Thermal Systems at Fraunhofer Institute for Solar Energy Systems ISE (2009–2014). Currently, he is pursuing a Ph.D. project that deals with the further development of performance test methods for multivalent heat transformers used for heating and cooling of buildings. He completed his Masters of Science in Renewable Energy and Energy Efficiency at University of Kassel in 2010. In 2008, he received his Diploma in Mechanical Engineering from Fachhochschule Bingen, University of Applied Sciences.

Wolfram Sparber is Head of the Institute for Renewable Energy at EURAC Research since its foundation in 2005. One of the main research areas of the Institute are sustainable heating and cooling systems with several projects brought forward in the field of solar thermal systems combined with thermally or electrically driven heat pumps. Since 2011, Wolfram Sparber is Vice President of the board of the European Technology Platform Renewable Heating and Cooling with a focus on hybrid systems including several heat sources in one system. As well in 2011 he took over the presidency of the board of SEL AG, a regional energy utility focused on renewable power production, energy distribution, and district heating.

List of contributors

Thomas Afjei
Fachhochschule Nordwestschweiz
(FHNW)
Institut Energie am Bau (IEBau)
St. Jakobs-Strasse 84
4132 Muttenz
Switzerland

Chris Bales
Solar Energy Research Center (SERC)
School of Industrial Technology and
Business Studies
Högskolan Dalarna
791 88 Falun
Sweden

Erik Bertram
Institut für Solarenergieforschung
Hameln (ISFH)
Am Ohrberg 1
31860 Emmerthal
Germany

Sebastian Bonk
University of Stuttgart
Institute for Thermodynamics and
Thermal Engineering (ITW)
Pfaffenwaldring 6
70550 Stuttgart
Germany

Jacques Bony
Haute Ecole d'Ingénierie et de Gestion du
Canton de Vaud
Laboratoire d'Energétique Solaire et de
Physique du Bâtiment (LESBAT)
Centre St-Roch
av. des Sports 20
1400 Yverdon-les-Bains
Switzerland

Sunliang Cao
Aalto University
School of Engineering
Department of Energy Technology
HVAC Technology
00076 Aalto
Finland

Daniel Carbonell
Hochschule für Technik HSR
Institut für Solartechnik SPF
Oberseestr. 10
8640 Rapperswil
Switzerland

Maria João Carvalho
Laboratório Nacional de Energia e
Geologia, I.P.
Laboratório de Energia Solar
Estrada do Paco do Lumiar 22
1649-038 Lisbon
Portugal

Matteo D'Antoni
EURAC Research
Institute for Renewable Energy
Viale Druso 1
39100 Bolzano
Italy

Ralf Dott
Fachhochschule Nordwestschweiz
(FHNW)
Institut Energie am Bau (IEBau)
St. Jakobs-Strasse 84
4132 Muttenz
Switzerland

Harald Drück
University of Stuttgart
Institute for Thermodynamics and
Thermal Engineering (ITW)
Pfaffenwaldring 6
70550 Stuttgart
Germany

Sara Eicher
Haute Ecole d'Ingénierie et de Gestion du
Canton de Vaud
Laboratoire d'Energétique Solaire et de
Physique du Bâtiment (LESBAT)
Centre St-Roch
av. des Sports 20
1400 Yverdon-les-Bains
Switzerland

Jorge Facão
Laboratório Nacional de Energia e
Geologia, I.P.
Laboratório de Energia Solar
Estrada do Paco do Lumiar 22
1649-038 Lisbon
Portugal

Roberto Fedrizzi
EURAC Research
Institute for Renewable Energy
Viale Druso 1
39100 Bolzano
Italy

Carolina de Sousa Fraga
University of Geneva
Institute for Environmental Sciences
Energy Group
Batelle Bat. D, Route de Drize 7
1227 Carouge
Switzerland

Robert Haberl
Hochschule für Technik HSR
Institut für Solartechnik SPF
Oberseestr. 10
8640 Rapperswil
Switzerland

Jean-Christophe Hadorn
BASE Consultants SA
8 rue du Nant
1207 Geneva
Switzerland

Michel Y. Haller
Hochschule für Technik HSR
Institut für Solartechnik SPF
Oberseestr. 10
8640 Rapperswil
Switzerland

Michael Hartl
AIT Austrian Institute of Technology
Energy Department
Giefinggasse 2
1210 Vienna
Austria

Andreas Heinz
Technische Universität Graz
Institut für Wärmetechnik (IWT)
Inffeldgasse 25/B
8010 Graz
Austria

Sebastian Herkel
Fraunhofer Institute for Solar Energy
Systems
Division of Thermal Systems and
Buildings
Heidenhofstr. 2
79110 Freiburg
Germany

List of contributors

Pierre Hollmuller
University of Geneva
Institute for Environmental Sciences
Energy Group
Batelle Bat. D, Route de Drize 7
1227 Carouge
Switzerland

Anja Loose
University of Stuttgart
Institute for Thermodynamics and
Thermal Engineering (ITW)
Pfaffenwaldring 6
70550 Stuttgart
Germany

Ivan Malenković
Fraunhofer Institute for Solar Energy
Systems
Division of Thermal Systems and
Buildings
Heidenhofstr. 2
79110 Freiburg
Germany
and
AIT Austrian Institute of Technology
Energy Department
Giefinggasse 2
1210 Vienna
Austria

Floriane Mermoud
University of Geneva
Institute for Environmental Sciences
Energy Group
Batelle Bat. D, Route de Drize 7
1227 Carouge
Switzerland

Marek Miara
Fraunhofer Institute for Solar Energy
Systems
Division of Thermal Systems and
Buildings
Heidenhofstr. 2
79110 Freiburg
Germany

Fabian Ochs
University of Innsbruck
Unit for Energy Efficient Buildings
Technikerstrasse 13
6020 Innsbruck
Austria

Peter Pärisch
Institut für Solarenergieforschung
Hameln (ISFH)
Am Ohrberg 1
31860 Emmerthal
Germany

Bengt Perers
Technical University of Denmark
DTU Civil Engineering DK & SERC
Sweden
Brovej, Building 118
2800 Kgs. Lyngby
Denmark

Jörn Ruschenburg
Fraunhofer Institute for Solar Energy
Systems
Division of Thermal Systems and
Buildings
Heidenhofstr. 2
79110 Freiburg
Germany

Christian Schmidt
Fraunhofer Institute for Solar Energy
Systems
Division of Thermal Systems and
Buildings
Heidenhofstr. 2
79110 Freiburg
Germany

Kai Siren
Aalto University
School of Engineering
Department of Energy Technology
HVAC Technology
00076 Aalto
Finland

Wolfram Sparber
EURAC Research
Institute for Renewable Energy
Viale Druso 1
39100 Bolzano
Italy

Bernard Thissen
Energie Solaire SA
Rue des Sablons 8
3960 Sierre
Switzerland

Alexander Thür
Universität Innsbruck
Institut für Konstruktion und
Materialwissenschaften
AB Enrgieeffizientes Bauen
Technikerstr. 19a
6020 Innsbruck
Austria

Martin Vukits
AEE – Institut für Nachhaltige
Technologien
Feldgasse 19
8200 Gleisdorf
Austria

IEA solar heating and cooling programme

The Solar Heating and Cooling Programme was founded in 1977 as one of the first multilateral technology initiatives ("Implementing Agreements") of the International Energy Agency. Its mission is to "advance international collaborative efforts for solar energy to reach the goal set in the vision of contributing 50% of the low temperature heating and cooling demand by 2030."

The member countries of the Programme collaborate on projects (referred to as "Tasks") in the field of research, development, and demonstration (RD&D) and test methods for solar thermal energy and solar buildings.

A total of 53 such projects have been initiated to date, 39 of which have been completed. Research topics include

- Solar Space Heating and Water Heating (Tasks 14, 19, 26, and 44)
- Solar Cooling (Tasks 25, 38, 48, and 53)
- Solar Heat or Industrial or Agricultural Processes (Tasks 29, 33, and 49)
- Solar District Heating (Tasks 7 and 45)
- Solar Buildings/Architecture/Urban Planning (Tasks 8, 11, 12, 13, 20, 22, 23, 28, 37, 40, 41, 47, 51, and 52)
- Solar Thermal & PV (Tasks 16 and 35)
- Daylighting/Lighting (Tasks 21, 31, and 50)
- Materials/Components for Solar Heating and Cooling (Tasks 2, 3, 6, 10, 18, 27, and 39)
- Standards, Certification, and Test Methods (Tasks 14, 24, 34, and 43)
- Resource Assessment (Tasks 1, 4, 5, 9, 17, 36, and 46)
- Storage of Solar Heat (Tasks 7, 32, and 42)

In addition to the project work, a number of special activities – Memorandum of Understanding with solar thermal trade organizations, statistics collection and analysis, conferences, and workshops – have been undertaken. An annual international conference on *Solar Heating and Cooling for Buildings and Industry* was launched in 2012. The first of these conferences, SHC2012, was held in San Francisco, CA.

Current members of the IEA SHC programme

Australia	Germany	RCREEE
Austria	Gulf Organization for Research and Development	Singapore
Belgium		South Africa
Canada		Spain
China	France	Sweden
Denmark	Italy	Switzerland
ECREEE	Mexico	The Netherlands
European Commission	Norway	Turkey
European Copper Institute	Portugal	United Kingdom

Further information

For up-to-date information on the IEA SHC work, including many free publications, please visit www.iea-shc.org.

Forewords

The steady growth of solar thermal systems for more than three decades has shown that solar heating systems are both mature and technically reliable. However, solar thermal systems have usually been sold as an add-on to a conventional hot water or space heating system.

In future, we need to develop hybrid systems that offer a complete heating system based on renewables, one that is able to cover 100% of the heating demand of buildings.

One very promising possibility is the combination of solar systems and heat pumps.

This book shows different ways in which these two technologies can be combined, and it presents the path to high-performance hybrid systems.

The book is the result of a collaborative international work within the Solar Heating and Cooling Programme and the Heat Pump Programme of the International Energy Agency (IEA). It is a great pleasure for me, as a former chairman of the IEA SHC Executive Committee, to introduce this book.

The international work has led to some interesting findings on solar and heat pump combinations based on monitored data and the use of simulation. The book presents all these findings and the methodologies needed to assess the energy performance of such combinations. It is an important contribution to the body of scientific knowledge on renewable heat that the IEA has been supporting for over 40 years.

I am sure that the reader will find new knowledge and inspiring ideas for future-oriented hybrid heating systems based on renewables.

Werner Weiss
IEA SHC Chairman 2010–2014

Achieving high-performance or net-zero-energy buildings in terms of energy consumption and greenhouse gas (GHG) emissions requires the utilization of energy-efficient technologies and renewable energy technologies. The combination of heat pumps and solar energy is a very promising solution for achieving this goal. This technology is what the participants of Annex 38 and Task 44 of the IEA Heat Pump Programme and the Solar Heating and Cooling Implementing Agreements, respectively, have together been exploring. The study results are presented in this book, which is a valuable reference and a significant contribution to HVAC renewable energy systems applications.

Sophie Hosatte
IEA HPP Chairman 2005–2014

Acknowledgments

All participants in T44A38, the four subtask leaders, and the Operating Agent would like to express their gratitude to the following for the given opportunity of undertaking the scientific and technical work in this project:

- The IEA Solar Heating and Cooling Programme Committee as well as the Heat Pump Programme Committee, and their secretariats, for accepting the theme of SHP as a research topic for an international collaboration, and partly financially supporting it.
- All national institutional bodies that have financially supported the work of every team, namely, the energy ministries of most participating countries.
- The industry partners that have worked with us and have understood that knowledge can be a necessary and durable asset for their business in the SHP field.
- All national partners that have been in contact with participants of the T44A38 and provided information on ongoing SHC projects or simulation tools.
- All owners of a house equipped with an SHP installation who have accepted the monitoring of their installation and the dissemination of results.
- The three reviewers of the T44A38 reports and book, Andréas Eckmanns from Switzerland, Ken Guthrie from Australia, and Michele Zinzi from Italy, all three experts representing their country in the SHC Executive Committee.

The Operating Agent wants to thank all participants in T44A38 for their active work and the four subtask leaders and their assistants who have committed themselves to organize this international project and to write the very technical subtask reports and most part of this book.

May 2015
Switzerland

Jean-Christophe Hadorn
Operating Agent

1 Introduction

Jean-Christophe Hadorn

1.1 The scope

This book is about a hybrid technology called "solar and heat pump." It is basically the combination of a solar system with a heat pump delivering heat to a building.

When the sun is shining, the collectors will be the primary source of energy for the domestic hot water preparation and for the space heating. Furthermore, the daily solar production can be stored for future use during a few days. When the sun is less abundant or when the solar storage is empty, the heat pump will take over the duty. The source of the primary low-energy "heat" for the heat pump to operate can be air, ground, or water from a river or an aquifer. A nice feature of the hybrid combination is that the solar collectors can also be used as the provider of the primary heat for the heat pump. The two components will then operate in the so-called serial mode.

This book will analyze the behavior of the main combinations of solar and heat pump, derive facts from practical projects, provide results from simulations and laboratory tests, and draw conclusions based on 4 years of activity of a collaborative project developed under the auspice of the International Energy Agency.

1.2 Who should read this book?

This book is recommended to the HVAC (heating, ventilation, and air conditioning) industry, the HVAC engineers and students, energy systems designers and planners, architects, energy politicians, manufacturers of solar energy components, manufacturers of heat pumps, standardization bodies, heating equipment distributors, and researchers in HVAC and building systems.

1.3 Why this book?

Producing heat from solar energy is an established technology since the 1990s.

The heat pump technology known since 1930 is becoming a standard solution to heat buildings and prepare domestic hot water in many countries. Both markets have shown growth since 2000, especially the heat pump market noticeably in countries with a high share of hydroelectricity in their energy supply.

For some years, systems that combine solar thermal technology and heat pumps have been marketed to provide space heating and to produce domestic hot water. The energy prices, the need to reduce the overall electricity demand or the strategy to move to more efficient solutions for heating than the current ones, the European Union legislation, and future scenarios calling for more renewable energies have driven the change.

A strong initial development of combination of solar and heat pump started some years ago with the help of early work from industry and research bodies in a few European countries. Innovative companies have shown success stories in this early period and continue to promote the advantages of solar and heat pump combinations based on real experience.

The IEA Solar Heating and Cooling Program (SHC) launched in 2010 a 4-year project called Task 44, "solar and heat pump systems." This was a joint effort with the IEA Heat Pump Program (HPP) under the name "Annex 38" to contribute to a better understanding of SHP (for "solar and heat pump") systems.

> This book presents the state of the art of the combined technology of solar collectors and electrical heat pumps based on the work undertaken within the Task 44 and Annex 38 project called T44A38 throughout this book. More than 50 participants from 13 countries have contributed to the collaborative effort over the 4 years of this international project.

It is anticipated that the electricity cost will increase on the planet in the future, due to CO_2 cost considerations and scarcity of energy resources. Solar photovoltaics might change the picture if the technology is massively adopted. But still, highly efficient heat pumps reducing the electricity demand will be needed to substitute the fossil heating solutions that dominate the world energy market in the 2010s. Combinations with solar collectors can increase the overall performance of a heat pump and will therefore also be an elegant solution of choice.

There are scientific and technological issues in integrating solar collectors and heat pump machines. The complexity lies in having two variable sources that should work together optimally. Heat storage management and control strategies are also of prime importance for optimal design. This book will present the challenges and some solutions on all aspects.

T44A38 has concentrated its efforts on electrically driven heat pumps, not because other techniques such as sorption machines are not possible but because no participants in this international activity presented a project with a thermally driven machine.

1.4   What you will learn reading this book?

This book deals with

- heating systems;
- heat distribution by water-based systems (e.g., radiators and floor heating systems);
- small-scale systems, one-family house to small dwellings (5–100 kW);
- electrically driven heat pumps;
- residential houses; and
- new buildings and buildings to be renovated.

1. You will understand that SHP systems are complex systems that need careful design and optimal integration.

1.4 What you will learn reading this book?

2. You will learn that good combinations of solar and heat pump can be achieved when the application is correctly done in adapted conditions. Examples are shown and discussed.
3. You will discover which definitions for the seasonal performance factor (SPF) of an installation are recommended for a comparison of heating solutions, for the assessment of technologies, for an environmental analysis, or for some economical considerations.
4. You will be able to classify all kinds of SHP systems in a new systematic way using the T44A38 "energy flow chart" diagram that you can use further to represent any kind of energy system.
5. You will see the advantages of different combinations of SHP found on the market and be able to challenge the vendor on the system design and performance.
6. You will learn that detailed simulation tools exist and a special framework to simulate SHP systems is available on the T44A38 web site.
7. You will understand which energy flows you should monitor in a project or in the laboratory and what SPF you can reasonably expect.
8. You will have the tools to assess the energy and economical benefits and other qualitative benefits that you can get out of a combination of SHP.
9. You will have a tool to evaluate the cost of the delivered heat of an SHP and what CO_2 reduction can be expected if you succeed to increase the SHP performance.

The logical organization of this book is the following:

– *Part One: Theoretical considerations.*
  Chapter 2 will tell you how the SHP systems found on the market in 2010–2012 can be classified in a systematic way thanks to the collaboration of more than 80 SHP companies.

  Chapter 3 explains the theory behind the main components that you will find in an SHP system: the collector, the storage tank, the borehole, and the heat pump.

  Chapter 4 presents all definitions of the performances of a complex system such as the coefficient of performance (COP) and all kinds of the SPF depending on the boundaries considered, and explains which should be used for which purpose.

  Chapter 5 reviews the methods to test SHP systems in a laboratory so that prototypes or commercial products can be characterized and even further optimized.

– *Part Two: Practical considerations.*
  Chapter 6 presents the basics of monitoring SHP systems and recommends the best practice in data acquisition of SHP systems. Results of relevant systems monitored *in situ* are presented.

  Chapter 7 shows how to simulate SHP systems with the T44A38 framework. It also presents important results on SHP combinations and sensitivity analyses. The benefit of a solar system in SHP is also quantitatively assessed through simulations.

  Chapter 8 provides an interesting approach to evaluate the cost of an SHP system that can be compared with a classical or non-solar system. What brings solar system to a heat pump is quantitatively evaluated and qualitative criteria are listed according to the numerous experts in T44A38 along the 4-year work.

Complementing this book, the web site of T44A38 provides all appendices. The simulation framework is, for example, downloadable as well as all other documents produced during the course of T44A38 and papers or contributions from T44A38 experts at scientific conferences.

Enjoy!

Internet sources

task44.iea-shc.org/publications

http://www.heatpumpcentre.org/en/projects/ongoingprojects/annex38

www.iea-shc.org/

www.heatpumpcentre.org/

www.ehpa.org/

www.estif.org/

www.rhc-platform.org

Part One
Theoretical Considerations

2 System description, categorization, and comparison

Jörn Ruschenburg and Sebastian Herkel

Summary

In the first step, several ways to analyze and categorize solar and heat pump (SHP) systems are presented. A graphical tool is introduced for visualizing the essence of different system concepts. The main criteria for analysis and categorization are found on component level (characteristics of solar collectors, heat pump, sizing, etc.) as well as on system level. For example, systems may cover space heating, domestic hot water (DHW) generation, and even space cooling – possibly but not necessarily more than just one of these functions. A system categorization approach is introduced – featuring parallel, series, and regenerative interactions between solar collectors and heat pump – that is applied throughout this book.

The precondition for defining performance figures and test methods for solar and heat pump systems (see Chapter 4) is a review of the market-available systems, investigating the relevance of nonstandard components and configurations. Such a review is presented, conducted on international level and followed by analysis regarding technical solutions on both component level and system level. Within this survey, carried out by IEA SHC Task 44/Annex 38 participants from several countries, 128 market-available solar and heat pump systems were identified. Most companies offer "conventional" systems with flat-plate collectors for both space heating and preparation of DHW. Still, manifold alternatives and technological as well as market-specific particularities are found. For example, solutions with photovoltaic–thermal collectors or heat pump solutions with solar thermal energy as only source exist since years ago.

2.1 System analysis and categorization

The aim of this chapter is to present possible ways to analyze and categorize existing and even future solar and heat pump systems. There are five main criteria to describe a solar and heat pump system: (i) the type of heat demand to be served; (ii) the low-temperature heat source(s) of the heat pump; (iii) the form(s) of energy used to drive the system; (iv) the function and placement of storages in the system; and (v) the interactions between these components. In addition, the systems can be described by the type of components used, by the sizing of components, and by the control. Thus, the reader should be aware that there is no global way of categorization that could meet all demands.

2.1.1 Approaches and principles

In the literature, various specifications are analyzed to describe or compare SHP systems [1]. Depending on the respective interests, independent authors focus on parameters that seldom coincide. The aspects chosen for the scope of this chapter and their usage by other authors are listed in Table 2.1.

Solar and Heat Pump Systems for Residential Buildings, First Edition.
Edited by Jean-Christophe Hadorn.
© 2015 Ernst & Sohn GmbH & Co. KG. Published 2015 by Ernst & Sohn GmbH & Co. KG.

Table 2.1 Examined parameters and their application in literature

Parameter	[2]	[3]	[4]	[5,6]	[1]
Provenance and distribution				x	
System functions	x	x		x	x
System concept	x	x	x		x
Heat pump characteristics		x	x	x	x
Collector characteristics		x			x

Provenance and distribution are more organizational than technical parameters, though there might be, for example, climatic design influences when comparing North and South European systems. System functions include DHW preparation as well as space heating and cooling. A system concept is usually defined by the interaction of heat pump, collector, and storages (cf. Section 2.1.3). Heat pump and collector characteristics may refer to several properties discussed in the respective subsections.

The possibilities to set up categories are equally varied. For example, the systems can be described by the type of applied components such as flat-plate, unglazed, or evacuated tube collectors, or alternatively by the refrigerant used in the heat pump cycle. The behavior of SHP systems depends also on the location, on the sizing of the components, and on the control. So, the definition of categories is multifaceted and strongly influenced by its purpose.

In this chapter, a graphical representation is introduced to systematically analyze and compare solar and heat pump systems. Afterward, these examinations result in a categorization approach.

2.1.2 Graphical representation of solar and heat pump systems

The visualization presented in this chapter, first published by Frank *et al.* [1], is similar to energy flow charts that are frequently used in building energy engineering. Instead of a whole building, it is the heating system that is illustrated centrally against white background, including energy-storing (blue objects) and energy-transforming components (orange objects). The analysis of many combined solar and heat pump systems resulted in the finding of five recurring components. They comprise collector, heat pump, and backup heater, complemented by storages, namely, one on the source side and one on the sink side of the heat pump. For these typical components, fixed positions are defined. Specifications such as the collector type may be chosen. As defined boundaries (gray background), environmental energy (green objects) enters the system from above, final or "to-be-purchased" energy – in case of electricity generating systems even "bidirectionally traded" energy – from the left side (dark gray objects), and useful energy such as DHW leaves to the right (red objects).

The information provided by the coloring is in any case an additional feature, that is, it is not essential for understanding. In theory, any heat losses would be shown leaving the system downward. However, because of the purely qualitative nature of the approach,

2.1 System analysis and categorization

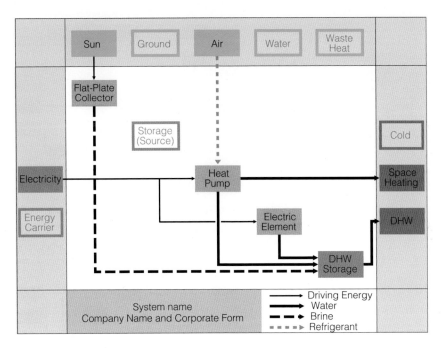

Fig. 2.1 Introductory example for the visualization scheme

component sizes, efficiencies, and so on are not shown, and thus no losses. A label for the manufacturer's and the concept's name is added in the lower left part.

The final step is the depiction of energy flows connecting certain components. In doing so, the figure is enhanced to become a qualitative energy flow chart. The line style refers to the carrier medium (water, brine, and refrigerant) or indicates driving energies such as solar irradiation, gas, or electricity. A simple example for a complete visualization is given by Figure 2.1.

It has to be pointed out that all possible operational modes of one system (excluding defrosting) are shown simultaneously in one scheme. All components appearing in a particular system are depicted as filled and nonexistent components remain as shaded frames as placeholders for orientation and comparability purposes. This arrangement results in energy flows mostly in left-to-right and up-down directions, though, of course, exceptions of this rule are possible.

In Figure 2.2, the presented visualization scheme is applied to typical systems. Here, a simplified hydraulic scheme is used as well, ignoring details such as backup heating elements. The comparison between the figure parts gives an impression of the visualization method's capability.

2.1.3 Categorization

System concepts are defined by the way the heat pump and the solar subsystem interact. Introductory works about the topic are found in Refs [7,8]. The distinctions made in this book, fully described in Ref. [9], are shown below.

10   2 System description, categorization, and comparison

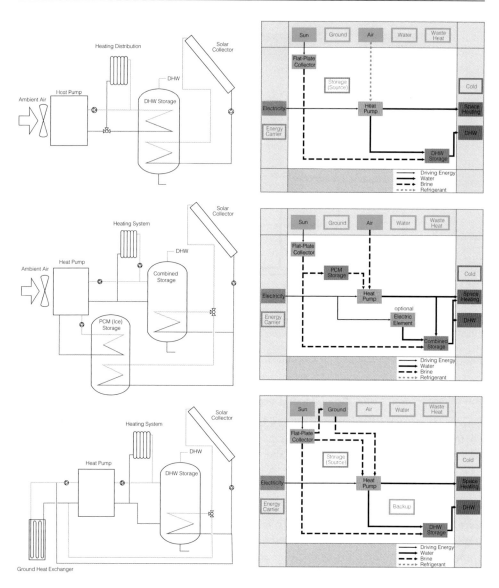

Fig. 2.2 Simplified hydraulic schemes (left) and corresponding visualizations (right) of different solar and heat pump systems

> Collector and heat pump independently supply useful energy (space heating and/or DHW), usually via one or more storages. This configuration is denoted as *parallel*, independent from the heat pump's source(s).
>
> The collector acts as a source of the heat pump, either as exclusive or as additional source, and either directly or via a buffer storage. This configuration is denoted as *series*.
>
> The use of solar energy to warm the main source of the heat pump, in this case usually ground, is denoted as *regenerative*.
>
> It is important to see that series and/or regenerative modes do not exclude each other within one system. Therefore, many system concepts are in fact combinations of these modes.

The regenerative approach – described in detail by Kjellson [10] and Meggers *et al.* [11] – could possibly be regarded as a subset of the series concept. There are conceptual and operational differences, though: The regenerative operation is usually applied to improve or at least to maintain the quality of the ground source for long timescales or merely to prevent solar collector stagnation. Consequently, regenerative operation usually occurs in summer, when highest solar availability and lowest heating demand concur, that is, when the heat pump is not in operation. Many systems were found on the market featuring a regenerative mode explicitly, partially without an intended series mode.

It has to be realized that parallel, series, and regenerative arrangements do not exclude each other within one system. Depending on the demand and the climatic conditions, system controllers and hydraulics may adopt more than just one of these approaches. The counting of all possible combinations – while ignoring permutations, redundancies, and the trivial case of "none" – results in seven options. The illustrative systems shown in Figure 2.2 shall be used to give examples. From top to bottom, a parallel, a parallel–series, and a parallel–series–regenerative system are shown. Cooling functions of SHP systems are not represented by this approach, though they might provide regenerative effects as well.

This approach with its seven categories can be applied to all solar and heat pump systems known today, as shown in the following sections.

2.2 Statistical analysis of market-available solar thermal and heat pump systems

Numerous combinations of heat pumps and solar thermal collectors became market-ready over the last few years. It is evident, as shown in Section 2.2.2.1, that some systems entered the market much earlier, motivated by the oil crisis around 1979. An explicit and enduring trend developed just in the current century, though.

Regarding testing and assessing SHP systems, existing methods and standards are limited (cf. Chapter 4). The use of solar thermal energy as a source for heat pumps, for example, is ignored by today's national and international standards. The precondition for defining performance figures and test methods for SHP systems is a review of the market-available systems, investigating the relevance of nonstandard components and configurations. Such a review is presented within this chapter, conducted on international level and followed by analysis regarding technical solutions on both component level and system level.

Earlier overviews of SHP systems were provided by Refs [2–5], identifying 5, 13, 19, and 25 systems, respectively. More recently, the work of Trojek and Augsten [5] was updated and confined to 19 systems by Berner [6]. A new approach – more international and more comprehensive – was launched within Task 44/Annex 38.

Intermediate results were presented by Ruschenburg *et al.* [9]. However, the companies analyzed at that time (September 2012) were rechecked in order to confine all analyses to those companies and products still existing and available in April 2014. All in all, the reviewed basis is now formed by 128 combined solar thermal and heat pump systems (minus 7), provided by 72 (also minus 7) companies from 11 countries. The methods and the results are presented in the following sections.

2.2.1 Methods

The presented and analyzed systems were surveyed between October 2011 and September 2012 by participants of T44A38. Companies were preferably searched and contacted by native speakers. Like all activities of T44A38, the market survey and the subsequent analyses are limited to SHP systems that are equipped with electrically driven compression heat pumps and designed for DHW preparation and/or residential space heating. Cooling functions are documented as supplementary information only.

In principle, any heat pump can be combined with any solar thermal collector. Therefore, only those companies were taken into account that genuinely provide at least one of the main components, that is, solar collector, heat pump, storage(s), and/or controller. Research projects were also ignored.

To ensure comparability, the characteristics of each SHP system were documented in a harmonized way on two-sided fact sheets, including data on the overall concept, hydraulics, dimensioning, and system control, as well as technical specifications mainly of the collector, heat pump, and storage(s). These documentations are found as online Annex on the T44A38 web page (task44.iea-shc.org).

Most data were derived from online or print sources, though personal contact to representatives of the companies could be established in most of the cases, enabling interviews. Anyhow, it is clear that the correctness of the retrieved information cannot be checked systematically and independently. Moreover, completeness cannot be claimed. The fact that the majority of identified companies originate from countries officially participating in T44A38 might possibly be explained by the barrier of language, resulting in certain (e.g., Asian) countries erroneously being underrepresented or even unrepresented.

It has to be noted that in the following analyses, all systems are treated equally, that is, without respecting the number of installations. Respecting the number of installations for each system would certainly lead to quite different results. The database is incomplete here, but when it comes to market penetration, most conventional approaches – those analogous to combinations of boilers and solar thermal collectors – outnumber the less classical configurations (cf. Section 2.2.2.3). Finally, it is pointed out that, due to imperfect data collection, the sample size is not necessarily constant throughout this

2.2 Statistical analysis of market-available solar thermal and heat pump systems

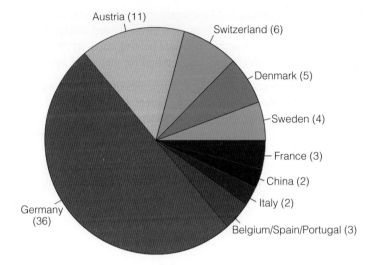

Fig. 2.3 Surveyed companies by country (country-code labeling according to ISO 3166-1)

chapter. From the 72 researched companies, for example, all appear in Figure 2.3 but only 56 in Figure 2.4. But as all figures are labeled with absolute numbers, the sample size can easily be calculated if desired.

Certain surveyed aspects will not be examined here. The reason is that their application turned out to be most flexible, and thus hardly comparable, comprising

– additional heat generators, that is, backup components such as electric heating elements, gas-fueled boilers, or wood stoves, sometimes offered optionally and almost always diversified regarding type, number, and way of integration;

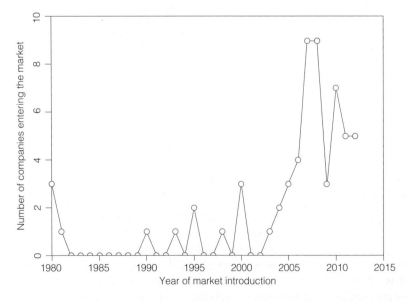

Fig. 2.4 Companies entering the market of solar and heat pump systems. (The oldest system offered by each company is used as the indicator, provided that it is still marketed today)

- storage characteristics, for example, number and function; separated storages for space heating and DHW can be chosen for most systems as well as combined storages; and
- the dimensioning of any components, for example, nominal heating capacity, collector area, and storage volume; these are too flexible to be specified for statistical analyses.

In the introduction of this chapter, it is implied that many companies offer more than one system. So, in conjunction with the discussion on parameters to be analyzed, the question arises which parameters are defined as being distinctive. Within the scope of the survey, either a different concept or a different source of the heat pump justifies a "distinguishable" system. This decision is to some extent arbitrary. Depending on the reader, the collector type or the refrigerant used within the heat pump might be regarded as more important, and a consideration of these aspects would unquestionably result in a multiple of "distinguishable" systems.

2.2.2 Results

2.2.2.1 Surveyed companies

As can be seen in Figure 2.3, most of the surveyed companies are based in Germany (50%) or Austria (15%). A complete list is found online on the T44A38 web site. Only a minority of all companies, however, restrict their market to one or two countries. The strong majority distributes their systems in three or more specified countries, even beyond those already named below, for example, to Croatia or to Greece. Being available "in Europe" or "worldwide" was claimed less frequently.

Figure 2.4 shows that most companies entered the SHP market in recent years. It has to be noted that systems withdrawn from the market – before, during, and after the time the survey was conducted – are ignored. About 15 are known to the authors.

2.2.2.2 System functions

The main functions for SHP systems, especially for residential applications, are space heating and the preparation of DHW. Figure 2.5 shows that both of these functions are featured in most cases. In contrast, few market-available systems are exclusively designed for DHW preparation. The fact that all Chinese systems and significant shares of the systems originating from the Mediterranean countries (France, Italy, and Spain) are included within the latter group is identified as a strong indicator for market-specific and climate-specific system layout. Regarding the technical design, these "DHW-only" systems can be divided into two groups: rooftop thermosyphon constructions backed up by an (air source) heat pump appear to be representative for China. In Europe, in contrast, storage and (often exhaust air) heat pumps are typically installed indoors as one integrated unit, with the condenser of the heat pump immersed in the storage tank or coiled around it.

Space cooling functions were surveyed supplementary. Interestingly, more than half of the systems (59%) are capable of "active" cooling via heat pump operation and/or "passive" cooling via ground or water source without heat pump operation, also known as "free," "natural," or "geo" cooling. As seen in Figure 2.5, this applies only to systems that already offer space heating. Some manufacturers offer all of their heat pump products with integrated cooling function by default.

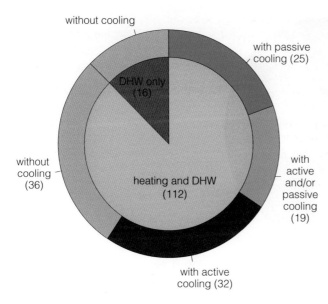

Fig. 2.5 Surveyed systems by function

It appears that air/air heat pumps, as popular as they might be for residential cooling and sometimes also heating purposes, are not combined with solar thermal systems. Instead, hydronic heat distribution is applied without exception. Here, floor heating systems are repeatedly recommended though rarely defined as mandatory.

As to the DHW preparation, modern hygienic approaches are found to be popular, given by internal heat exchangers (20% of the systems), for example, corrugated pipes, by external fresh water stations (23%), or by either solution to be selected (additional 6%). Austrian companies offer such technologies much above average.

2.2.2.3 System concepts

At this point, the system concepts as described in Section 2.1.3 are applied to the reviewed systems. The result is a fragmentation shown in Figure 2.6.

The "parallel-only" concept, which is simpler in design, installation, and control, clearly dominates (63%). SHP systems with "series-only" concepts (6%) or "regenerative-only" concepts (1%) are rare. Most impressively, concepts with any combination of parallel, series, and/or regenerative modes amount to no less than 30%.

2.2.2.4 Heat pump characteristics – heat sources

Leaving aside air/air heat pumps mainly used for space cooling in Mediterranean countries, it can be said that ambient air and ground are the most common sources for heat pump installations in Europe, while water and exhaust air cover smaller shares [12]. Though not recorded by such statistics, Figure 2.6 demonstrates that energy converted by solar collectors is repeatedly utilized as a source, that is, the series concept and its

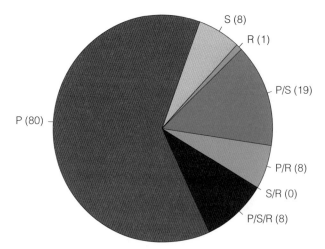

Fig. 2.6 Surveyed systems by concept (P: parallel; S: series; R: regenerative)

variations. It becomes clear that even the "series-only" concept allows other possible sources within the same system.

Regarding the classical sources, Figure 2.7 illustrates that either pure air source or pure ground source heat pumps together are applied in half of the surveyed systems, namely, 27 and 24%, respectively. Water (9%) or exhaust air (6%) are utilized as a source in a few systems. Commercial SHP systems with wastewater or other sources appear untraceable.

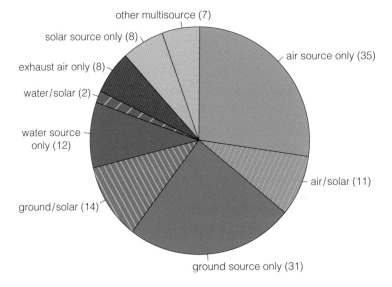

Fig. 2.7 Surveyed systems by source

Systems using solar energy as the sole source amount to 6%. For a further 21%, solar energy is used in addition to other conventional sources (air, ground, or water). Such multisource systems require technical solutions that can be split into two groups, external and internal ones. The former refers to modified hydraulics between solar subsystem and heat pump, for example, by heat exchangers between solar loop and ground source loop or even by joint brine loops. Thus, conventional heat pumps and solar collectors can be used. The latter means that either the heat pump or the solar collectors are specifically designed to be integrated within series SHP systems. Though this approach is rare, offered by not more than six companies, it comprises the most alternative solutions, including

- multisource evaporators (two evaporators within one refrigerant cycle);
- directly evaporating solar collectors (with refrigerant as circulating fluid for the solar loop); and
- hybrid collectors (the solar thermal collectors include also the ambient air unit with integrated fan or other "active" technology).

2.2.2.5 Collector types

Within the conducted survey, the collector type was chosen as the most significant parameter to compare the applied solar subsystem. The results are shown in Figure 2.8. Questions on additional characteristics – for example, regarding the circulating fluid, material, operational modes, Solar Keymark certification, or handling of stagnation – were answered incomprehensively; thus, these parameters cannot be presented in comparative form.

Flat-plate collectors (FPCs) are stipulated in nearly half of the systems (48%), whereas evacuated tube collectors (ETCs) are essential only in the fewest cases (2%). Instead, the choice between these two types is frequently left open, that is, affected by the conditions

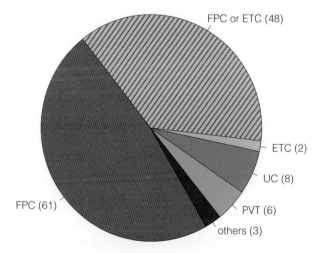

Fig. 2.8 Surveyed systems by collector type

on site as well as the preferences of client and installer (38%). Uncovered or unglazed collectors (UCs) are found repeatedly (6%), mainly in specific applications (cf. Section 2.2.2.6). Recently developed PVT collectors are found only in few market-available SHP systems (5%), according to this survey for the first time in 2011.

2.2.2.6 Cross analysis between collector type and system concept

When compared with flat-plate and evacuated tube collectors, photovoltaic–thermal and unglazed collectors are efficient only at low temperatures and thus inefficient for space heating and even more when temperatures sufficient for DHW preparation are to be met. The source temperature of heat pumps is by definition too low to be used directly for heating purposes, though desired to be as high as possible – given the conditions at site – to increase the heat pump's efficiency. The logical reasoning out of these observations would be that, if installed within SHP systems, UCs and PVT collectors are preferably applied to concepts with series and/or regenerative character while FPCs and ETCs are favored for parallel concepts.

The acquired data (cf. Figures 2.6 and 2.8) allow verification by means of correlation. Figure 2.9 is a bubble plot with discrete values in all dimensions. The area of each bubble is proportional to the number of respective systems.

FPCs and ETCs preponderate within the "parallel-only" concept and all of its combinations. In contrast, UCs and PVT collectors are never applied to "parallel-only" systems but represent most of the "series-only" systems and the single "regenerative-only" product. This is especially interesting given the fact that FPCs or ETCs are found in 87% of the systems in total. All in all, it is evident that the marketed systems mirror the theoretical considerations given above.

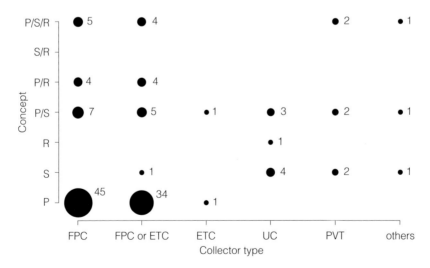

Fig. 2.9 Surveyed systems correlated by collector type and system concept (P: parallel; S: series; R: regenerative; FPC: flat-plate collector; ETC: evacuated tube collector; UC: unglazed collector; PVT: photovoltaic–thermal collector)

2.3 Conclusions and outlook

Within a market survey carried out by participants from several countries, 128 market-available SHP systems were identified. Most companies offer "conventional" systems with FPCs in parallel arrangement for both space heating and preparation of DHW. Still, manifold alternatives are found for each analyzed parameter.

As a main result of this survey, many technological and market-specific particularities of SHP systems are identified. For example, the fact that all Chinese systems and significant shares of the systems originating from the Mediterranean countries are designed for DHW preparation without space heating is identified as strong indicator for market-specific and climate-specific system layout.

Probably the best example regarding the system design is the extensive usage especially of uncovered collectors in series concepts. Here, it is evident that the industry follows theoretical considerations in both ecology and economy. This approach is, by definition, infeasible for non-solar and heat pump systems as well as for solar heating systems combined with any fossil fuel heating.

As additional result of this survey, the need for flexible performance figures and testing methods becomes visible. It was, for example, shown that series or more complex configurations are widely distributed in Europe, which is ignored by today's standards. In these cases, solar energy is extensively used as source energy for the heat pump though not directly as useful energy. The application of PVT collectors in market-ready systems is a rather young trend. Consequently, assessment methods are required taking into account the energy both consumed from and fed into the grid, similar to cogeneration systems.

Based on these results, performance figures as well as monitoring concepts can be developed to assess existing and even future systems (cf. Chapter 4). For example, it may be found that certain concepts are in general less efficient or that increased technical complexity required for some configurations does not result in a corresponding increased performance.

The trend shown in Figure 2.4 suggests that further companies will enter the market of SHP systems in recent years. Furthermore, even systems offered today may be subject to changes in system concept or its components. The introduction of PVT collectors and the choice of refrigerants are examples. The dynamic history should remind the readers that the situation as presented in this chapter is merely a snapshot.

2.4 Relevance and market penetration – illustrated with the example of Germany

This section deals with the question to what extent SHP already penetrated the market. Results can be stated either in absolute numbers or relative to conventional systems, that is, non-solar and heat pump systems.

Unfortunately, such numbers are not monitored systematically, either on national or on international level. Only separated numbers, without information about possible combinations, are well established for both heat pumps and solar thermal collectors.

For example, the European Solar Thermal Industry Federation (ESTIF) covers European solar thermal markets in great detail, and its comprehensive statistics are publicly available online. Similarly, the European Heat Pump Association (EHPA) presents the development of European heat pump markets, though for members only. However, neither of these or other institutions deals with combination.

Furthermore, the manufacturers of heating equipment present their numbers carefully. In personal communications to the authors, they typically state that between 5 and 10% of the produced heat pumps are installed with a solar thermal combination. A company distributing products as separated units instead of system solution is usually not even able to provide such numbers. Partly, companies are not willing to state any numbers for business reasons.

Nevertheless, some other indicative values can be presented here, only referring to the German market. There are two reasons why Germany is chosen here. First, Figure 2.3 shows that the German market is most considerable compared with any other monitored country. Second, it is the only country where data were available at all, aggregated by four independent institutions. The results are explained in the following:

- *The Federal Statistical Office of Germany (Statistisches Bundesamt):* This federal authority started recently to evaluate building permissions regarding not only primary heating technology but also additional components. Thirty-six thousand one hundred sixty heat pumps are listed among the permissions for new residential buildings in 2012. Among these are 2610 solar thermal combinations, or 7.2% in relative terms (unpublished data/personal communication).
- *The Federal Office of Economics and Export Control (Bundesamt für Wirtschaft und Ausfuhrkontrolle (BAFA)):* Among other activities, this office promotes renewable energies by means of subsidization. If a heat pump is installed in combination with a solar thermal system, an additional "combination bonus" can be granted, introduced in 2008. Since then, about 11% of the supported heat pump installations also received the combination bonus [13]. It has to be noted that BAFA subsidies are only applicable to renovation projects of certain size, and only upon application. Thus, less than a quarter of the German heat pump or solar thermal market is covered.
- *Institut Wohnen und Umwelt (IWU) and Bremer Energie Institut (BEI):* These institutes used a method where 7500 questionnaires were answered by building owners or property managers in 2009. After evaluating and scaling up of these data, it is found that heat pumps are installed in about 1.5% of about 18 million residential buildings in Germany. SHP systems exist in about 0.4% of the buildings; in other words, 26% of the heat pumps are installed in a solar thermal combination. Information regarding the statistical uncertainties can be found in the report [14].
- *The Institute for Heating and Oil Technology (Institut für Wärme und Oeltechnik (IWO)):* This institute conducts a yearly survey with about 1000 heating installation companies, asking for the number of installations of typical heat generators, and for the combination with solar thermal energy in particular. Both new buildings and renovation projects are covered. According to these surveys between 2008 and 2011, about 20 and 27% of the electrically driven heat pumps were combined with solar thermal systems [15].

All these publications cover not only heat pumps but also other heat generators that can be combined with solar thermal systems. For the data available from the Federal

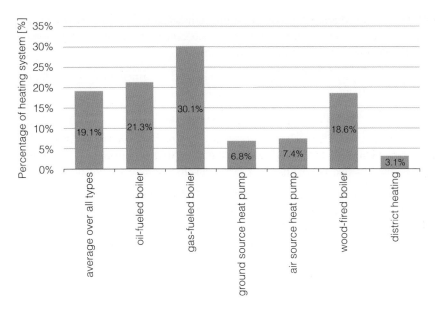

Fig. 2.10 Percentage of individual types of heating systems (space heating and DHW) for residential buildings that are combined with solar thermal systems. The left-hand bar gives the analogous average over all types of heating systems (based on building permissions for new residential buildings in Germany 2012, data from Federal Statistical Office of Germany)

Statistical Office of Germany, Figure 2.10 shows that the share of combined heat pumps is low when compared with gas, oil, wood, and consequently with the residential sector in general. At least to some extent, these differences can be explained by German building regulations demanding a renewable energy share in new building heating systems. Contrary to heat pumps, which are accepted as partly renewable even without solar components, it is more common to combine oil or gas with solar thermal energy to meet the requirements. Not presented in Figure 2.10 is the influence of building size. In houses for one or two families, 7% of heat pumps (all sources) are combined with solar thermal collectors, but more than 11% when examining houses for three or more families.

References

1. Frank, E., Haller, M., Herkel, S., and Ruschenburg, J. (2010) Systematic classification of combined solar thermal and heat pump systems. Proceedings of the EuroSun International Conference on Solar Heating, Cooling and Buildings, September 28–October 1, Graz, Austria.

2. Tepe, R. and Rönnelid, M. (2002) Solfångere och värmepump: Marknadsöversikt och preliminära simuleringsresultat, Centrum för Solenergiforskning, Solar Energy Research Center, Högskolan Dalarna, Borlänge, Sweden.

3. Müller, H., Trinkl, C., and Zörner, W. (2008) Kurzstudie Niederst- und Niedertemperaturkollektoren, Hochschule Ingolstadt, Germany.

4. Henning, H.-M. and Miara, M. (2009) Kombination Solarthermie und Wärmepumpe – Lösungsansätze, Chancen und Grenzen. Proceedings of the 19th Symposium "Thermische Solarenergie", Bad Staffelstein, Germany.

5. Trojek, S. and Augsten, E. (2009) Solartechnik und Wärmepumpe – sie finden zusammen. *Sonne Wind & Wärme*, **33** (6), 62–71.

6. Berner, J. (2011) Wärmepumpe und solar – solarenergie den vortritt lassen. *Sonne Wind & Wärme*, **35** (8), 182–186.

7. Freeman, T.L., Mitchell, J.W., and Audit, T.E. (1979) Performance of combined solar-heat pump systems. *Solar Energy*, **22** (2), 125–135.

8. Citherlet, S., Bony, J., and Nguyen, B. (2008) SOL-PAC: Analyse des performances du couplage d'une pompe à chaleur avec une installation solaire thermique pour la renovation. Final report, Haute Ecole d'Ingénierie et de Gestion du Canton de Vaud (HEIG-VD), Laboratoire d'Energétique Solaire et de Physique du Bâtiment (LESBAT), Yverdon-les-Bains, Switzerland.

9. Ruschenburg, J., Herkel, S., and Henning, H.-M. (2013) A statistical analysis on market-available solar thermal heat pump systems. *Solar Energy*, **95**, 79–89.

10. Kjellson, E. (2009) Solar collectors combined with ground-source heat pumps in dwellings. Ph.D. thesis, Lund University, Sweden.

11. Meggers, F., Ritter, V., Goffin, P., Baetschmann, M., and Leibundgut, H. (2012) Low exergy building systems implementation. *Energy*, **41**, 48–55.

12. Nowak, T. and Murphy, P. (2012) Outlook 2012 – European Heat Pump Statistics, European Heat Pump Association, Brussels, Belgium.

13. Bundesamt für Wirtschaft und Ausfuhrkontrolle (BAFA) (2012) Marktanreizprogramm – geförderte Anlagen.

14. Diefenbach, N., Cischinsky, H., Rodenfels, M. (Institut Wohnen und Umwelt), and Clausnitzer, K.-D. (Bremer Energie Institut) (2010) Datenbasis Gebäudebestand – Datenerhebung zur energetischen Qualität und zu den Modernisierungstrends im deutschen Wohngebäudebestand, Darmstadt, Germany.

15. Institut für Wärme und Oeltechnik (IWO) (2012) Anlagenbaubefragung 2008–2011 – Solaranteile in Modernisierung plus Neubau, Hamburg, Germany.

3 Components and thermodynamic aspects

Michel Y. Haller, Erik Bertram, Ralf Dott, Thomas Afjei, Daniel Carbonell, Fabian Ochs, Andreas Heinz, Sunliang Cao, and Kai Siren

> **Summary**
>
> The main components of solar and heat pump (SHP) systems are introduced. These are the solar collectors (Section 3.1), the heat pumps (Section 3.2), ground source heat exchangers (Section 3.3), and thermal energy storage (Section 3.4).
>
> These components are discussed from a technical point of view, explaining the main factors that influence their energetic performance and characteristics that may be of importance for the particular application in combined solar and heat pump systems. The complexity of solar and heat pump systems often requires energetic performance simulation in order to obtain reliable estimations for the performance of the overall systems. Therefore, the thermodynamic background for the simulation of each component is introduced, and a selection of simulation models is presented. The focus is thereby clearly on models and model features that are of particular interest for simulations of combined solar and heat pump systems.
>
> It is important to note that in series and in regenerative SHP systems, the operating conditions of the components may be outside of the usual operating conditions in parallel SHP systems or in other systems that include solar thermal or heat pumps. This may bring along new requirements for the design of these components; for example, selective absorber coatings and collector insulation must be compatible with condensation on the absorber surface, or the refrigerant cycle of a heat pump must be designed differently in order to benefit from higher source temperatures. This may influence the choice of the compressor and the expansion valve. It also means that existing component simulation models must be validated and possibly adapted for the new operating conditions that they are used for in SHP systems.
>
> Finally, Section 3.5 presents special aspects concerning the combination of solar collectors and heat pumps. These are, for example, the question whether and when to use heat from solar collectors for the evaporator of the heat pump instead of using it in parallel to the heat pump operation, and the importance of exergetic efficiency and storage stratification for SHP systems.

3.1 Solar collectors

Solar thermal collectors convert solar radiation to usable heat. At present, the vast majority of the solar collectors in combination with heat pumps are nonconcentrating, liquid-cooled, thermal collectors. Thus, the most distinguishing attribute of the collectors is whether the absorber is covered by a transparent cover or not. The performance characteristic of covered and unglazed collectors is obviously quite different (Figure 3.1a).

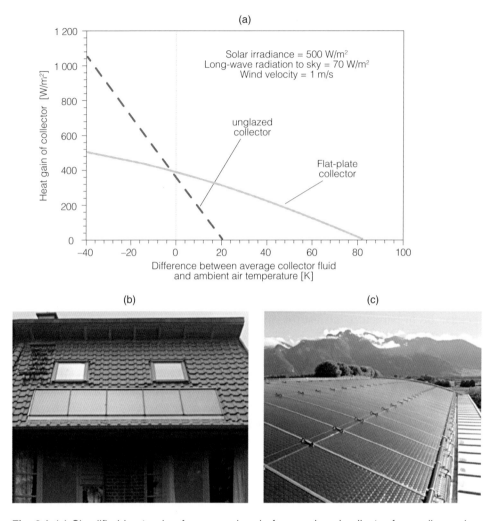

Fig. 3.1 (a) Simplified heat gain of a covered and of an unglazed collector for medium solar irradiance of 500 W/m^2; (b) covered thermal collector; and (c) selective unglazed collectors in Vouvry, Switzerland

In addition, heat pump systems allow the operation of unglazed absorbers as air heat exchangers even in the absence of solar irradiance and at nighttime.

Common design examples for covered solar thermal collectors are flat-plate collectors (Figure 3.1b) and vacuum tube collectors. Their transparent cover reduces the thermal convective exchange, which allows for a higher efficiency at higher temperature levels compared with unglazed collectors. Accordingly, they are better suited to provide heat for domestic hot water (DHW) and space heating directly, without the need of a heat pump.

3.1 Solar collectors

An unglazed absorber (Figure 3.1c) is the most basic solar thermal collector design. Due to the absence of a transparent cover, the thermal performance is characterized by larger solar gains at temperature levels close to the ambient air, but also by higher convective heat exchange with the ambient air and thus smaller gains at higher temperatures. Their typical application is swimming pool heating. On the other hand, for operation below the ambient air temperature the high convective heat exchange significantly improves the collector performance compared with the covered design.

Among emerging technologies are photovoltaic–thermal (PVT) collectors that may again be designed with or without an air gap between the transparent cover and the PV cells or the thermal absorber. Especially in combination with heat pumps, these PVT collectors are estimated to have a high potential and a fast-growing market. In principle, their thermal performance is lower than the performance of standard thermal collectors because of the conversion of part of the solar irradiation into electricity and due to compromises in the collector design for the integration of the PV cells. The PV performance increases slightly compared with a panel with PV cells if the operating temperatures of the PV cells are reduced due to the cooling effect of the heat extraction. This is typically the case for unglazed designs, but it may not be the case for covered designs.

In combination with a heat pump, solar collectors may be operated below the temperature of the ambient air and its dew point. Classical collector energy balances have not been designed for this operation mode. Therefore, they may have to be adapted in order to account correctly for the heat flows that appear under low-temperature conditions. A general energy balance including all possible energy flows is presented in Figure 3.2 and Equation 3.1.

The possible heat gains consist of the absorbed shortwave solar radiation $\dot{q}_{rad,S}$, longwave radiation exchange $\dot{q}_{rad,L}$, convective heat exchange with the air (split into sensible heat exchange $\dot{q}_{amb,sens}$ and latent heat exchange $\dot{q}_{amb,lat}$), heat conduction $\dot{q}_k$ (usually at the rear side), and energy gains from rain $\dot{q}_{rain}$. The latent heat exchange may be further split into the condensation $\dot{q}_{amb,cond} > 0$, evaporation $\dot{q}_{amb,cond} < 0$, frost

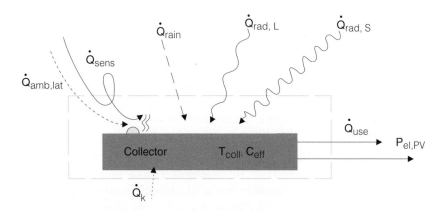

Fig. 3.2 Energy balance of a PVT collector or absorber

formation $\dot{q}_{\text{amb,frost}} < 0$, and frost melting $\dot{q}_{\text{amb,frost}} > 0$. In general, all these terms may appear not only on the front but also on the rear side of the absorber.

$$\frac{\dot{Q}_{\text{gain}}}{A_{\text{coll}}} = \dot{q}_{\text{gain}} = \dot{q}_{\text{rad,S}} + \dot{q}_{\text{rad,L}} + \dot{q}_{\text{amb,sens}} + \dot{q}_{\text{amb,lat}} + \dot{q}_{\text{k}} + \dot{q}_{\text{rain}}, \qquad (3.1)$$

with $\dot{q}_{\text{amb,lat}} = \dot{q}_{\text{amb,cond}} + \dot{q}_{\text{amb,frost}}$.

The useful power output of the collector consists of the usable heat $\dot{q}_{\text{gain}}$ and, in case of PVT collectors, additionally of electric power $p_{\text{el,pv}}$. Moreover, it accounts for the internal energy change by the effective thermal capacitance c_{eff} of the collector (Equation 3.2):

$$\dot{q}_{\text{use}} = \dot{q}_{\text{gain}} - p_{\text{el,pv}} - \frac{\delta T_{\text{coll}}}{\delta \tau} c_{\text{eff}}, \qquad (3.2)$$

where $\delta T_{\text{coll}}/\delta \tau$ is the time derivative of the average temperature of the thermal capacitance of the collector. Incoming heat flows are counted positive and outgoing flows are counted negative in accordance with Figure 3.2.

Collector simulation models provide a solution for the energy balance presented in Equations 3.1 and 3.2. Most of them are neglecting or simplifying one or more of the heat transfer mechanisms.

The combination of solar collectors with a heat pump may lead to an extension of the typical operating range including periods without solar irradiance and with collector temperatures that are below the ambient air temperature. In these cases, three additional effects may have an impact on the performance of the solar thermal collectors:

– Condensation
– Freezing
– Heat gains from rain

Out of these effects, energy yields from condensation seem to be the most relevant ones, at least for unglazed collectors. In addition, a correct prediction of the energy yield at nighttime operation without solar irradiance is important for these applications.

Condensation heat gains have been included in several collector models reported in the literature [1–9]. A common feature of these models is that the condensation heat gain is based on the theory of heat and mass transfer as presented in standard textbooks. Usually the model equations are based on a convective heat transfer coefficient, the relative humidity of the ambient air, the phase change enthalpy of water, and the difference between the water vapor load of the ambient air and the water vapor load at the surface of the absorber. However, the models differ in the assumption of the temperature at which the maximum water vapor load at the absorber surface is evaluated.

Validation of recent unglazed collector models with condensation has been done by several authors [6,10,11]. Eisenmann et al. [6] measured a metal roof collector in a laboratory wind channel for different humidity, low inlet temperatures, and wind speeds between 1.5 and 3 m/s. Perers et al. [10] measured a polymeric absorber with high thermal conductivity under outdoor conditions. Within T44A38, Bunea et al. [11]

compared both models under Swiss weather conditions with 5 days of field measurements of a selective, rear side insulated, unglazed collector.

The maximum error of the daily yield was determined to be approximately 50% for condensation gains [6,11], although the deviations were much smaller under most operation conditions and are not detectable for longer operation periods. The high deviations under certain conditions are explained by the simplifying assumption of constant values for surface emissivity, internal thermal heat conductivity, or convective heat loss coefficient of the absorber. In fact, the models displayed excellent accordance to reality for outdoor measurements and longer periods that include condensation and nighttime operation [10,11]. To conclude, high model errors for condensation can occur for particular conditions, because of the difficulty in determining a universal applicable parameter setting. On the other hand, all conducted outdoor measurements over longer periods and including condensation show very good model accordance with measurements.

The influence of wind on the convective heat and mass transfer coefficients of the absorber is an often discussed topic with large uncertainties. A wide range of different models exists both for the estimation of local wind speed based on meteorological wind speed and for the estimation of the effect of local wind speed on the convective heat transfer. A review of wind convection coefficient correlations has been presented by Palyvos [12]. Theoretically, in the absence of wind, the natural convection heat transfer coefficient of a cooled plate depends on its inclination. Although Philippen et al. [13] measured increased heat gains for larger inclination of the absorber under outdoor conditions, these increased gains were caused by higher long-wave irradiance from the field of view of the absorber. A dependence of the convective heat transfer coefficient on the inclination of the absorbers was not detected.

For system simulations, performance models based on standard test data are used for different platforms such as TRNSYS, MATLAB, or IDA ICE. They allow the assessment of different systems and configurations. The model choice depends strongly on the collector application in the system and on the aim of the investigation. Apart from this general recommendation, the following aspects should be kept in mind in order to ease the model choice and reliable use:

For covered collectors
- Covered collectors are not likely to get any significant energy yields from latent heat exchange. Hence, no latent heat gains have to be considered in the simulation model. Moreover, for most of these collectors condensation inside the collector may damage the selective coating or the insulation and thus has to be avoided.
- The quadratic loss term a_2 of the collector equation according to the European standard EN 12975 suggests increasing heat losses irrespective of the sign of the temperature difference between the collector and the ambient air. For unglazed collectors, it is generally recommended to set $a_2 = 0$. For other collectors that are operated below the temperature of the ambient air, the term $a_2 \cdot \Delta T^2$ may be replaced with $a_2 \cdot \Delta T \cdot |\Delta T|$.

For unglazed collectors
- For the special case of unglazed solar collectors with selective surfaces, dew on the absorber instantly changes its optical properties. In such a case, the model parameters may have to be adjusted in the course of the simulation.

- The definition of an efficiency based on a division of the heat output by the solar irradiance is not meaningful for the operation of unglazed collectors as ambient air heat exchangers in the absence of solar irradiance. Models that rely on such a calculation may produce unexpected or wrong results.
- For small mass flow rates and in the absence of solar irradiance, the simplifying assumption of a linear increase of the temperature of the fluid between the inlet and the outlet that is often made (e.g., in European standards) has to be checked carefully. A different approach may have to be chosen or the thermal capacitance of the collector may be discretized into several control volumes along the fluid flow path.

For covered and unglazed PVT collectors
- For the simulation of photovoltaic cells with thermal absorbers, the so-called PVT collectors, a subtraction of the photovoltaic yield from the available solar radiation must be applied to common thermal collector models.

Concluding, it can be said that under the special conditions of solar collector operation as a heat source for heat pumps, conventional collector simulation models may produce unexpected results. In particular, special attention should be paid to the operation without solar irradiance, below ambient air temperature, and with extremely low mass flow rates.

Table 3.1 lists some of the frequently used simulation models for solar collectors in combination with heat pump systems. Examples for apparent model limits are heat gains from freezing water, effects of rain, or collectors with forced air convection. Consequently, new models and extensions have to be developed continuously.

3.2 Heat pumps

Heat pumps are able to extract heat from the environment at a low temperature level and figuratively spoken "pump heat" to a higher temperature level. In order to drive this "heat pumping process", an energy source with higher exergy content is needed. In the case of electrically driven compression heat pumps that are in the focus of this chapter, the higher exergy source is electricity that is used to drive the compressor of the refrigerant cycle. Figure 3.3 shows the concept of a heat pumping cycle: (a) schematic, (b) in terms of a Sankey diagram, and (c) an example for the thermodynamic states of the refrigerant, together with the enthalpies and temperatures of the brine source and the heated water. The discussion of the heat pumping process in detail is beyond the scope of this chapter, and the reader should refer to general textbooks on the matter.

The energy balance of a (loss-free) heat pump process can be written as

$$\dot{Q}_{sink} = \dot{Q}_{source} + P_{el,comp}, \tag{3.3}$$

where $\dot{Q}_{sink}$ is the useful heat output from the condenser, $\dot{Q}_{source}$ is the heat input into the evaporator, and $P_{el,comp}$ is the electric power needed to drive the compressor. The coefficient of performance (COP) is defined as the amount of useful heat divided by the electric input:

$$COP = \dot{Q}_{sink}/W_{el}. \tag{3.4}$$

3.2 Heat pumps

Table 3.1 Collector models for solar and heat pump systems (nonexhaustive)

Platform – name and ID	Type of collector	Type of model	Heat exchange effects[a]				Comments	Documentation/ validation references
			Wind	Condensation	IR	Capacities		
TRNSYS Type 132	CC, UC	g	√	—	√	1×2		D: [14]
TRNSYS Type 136	CC, UC	g	√	√	√	1×1	Based on Type 132	D/V: [9,14]
TRNSYS Type 202/203	UC	g	√	√	√	$N\times 1$	Unglazed collector or PVT	D/V: [6,8]
TRNSYS Type 222	UC	w	√	√	√	1×1	Including rear side losses	D: [7,15]
TRNSYS Type 301	CC	g	√	—	√	$N\times 1$		D: [16]
TRNSYS Type 832 v5.00	CC, UC	g	√	√	√	$N\times 1$	Based on Type 132	D: [14,17]
RDmes white1	CC	w	√	—	√	$N\times 1$	IAM from optical properties	D: [18]
Matlab_Carnot	CC	w	No information available			10×1		D: [16]
T-Sol	CC, UC	g	No information available					D: T Sol User Manual
Polysun	CC, UC	g	√	—	√	1×1	Standards: USA, EU, China	D: [19] and collector test standards

CC = covered collector (evacuated tube/flat plate collector); UC = unglazed collector; w = white box (physical) model; g = gray box (semi-empirical model); IAM = incident angle modifier approaches: b_0/b_1 = first- or second-order IAM ASHRAE ([20], p. 298); r = Ambrosetti-r [21]; biax = also biaxial IAM calculation possible; table: IAM can be read in from a performance map table/data file; wind = effect of wind speed; cond = condensation when operated below dew point; IR = infrared radiation balance; capacities = number of heat capacity nodes along the fluid path times number of heat capacities from fluid to ambient temperature; N = variable number of nodes; D = documentation only; V = validation.
[a] Beam and diffuse radiation is considered for all models.

The theoretical upper limit for the COP is defined by the Carnot efficiency of the reversible heat pump cycle, that is,

$$\text{COP}_{\text{lim}} = T_{\text{sink}}/(T_{\text{sink}} - T_{\text{source}}), \quad (3.5)$$

where the temperatures T for the sink and the source are to be applied in K.

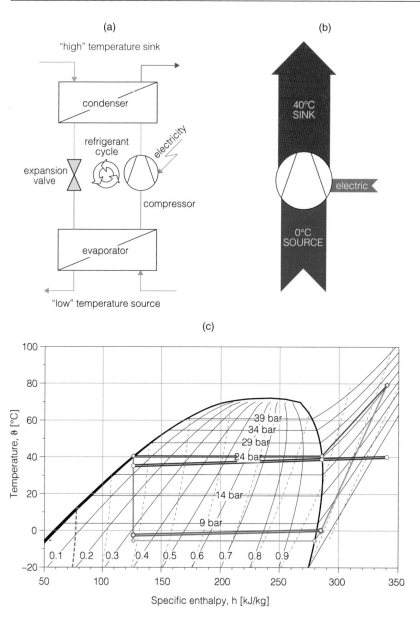

Fig. 3.3 Heat pump cycle: (a) schematic; (b) Sankey diagram; (c) thermodynamic states of a refrigerant with source and sink temperatures. (*Source*: Andreas Heinz, TU Graz and Michel Haller, SPF.)

For the application in dwellings, commonly used heat sources today are the outdoor air or the ground. In an air source heat pump, the evaporation usually takes place directly in an air-to-refrigerant heat exchanger that may be located within the envelope of the building or outside the building (split units or outdoor installed monoblocks as shown in Figure 3.4a). In most ground source heat pumps, a brine circuit transports the heat from the ground to the

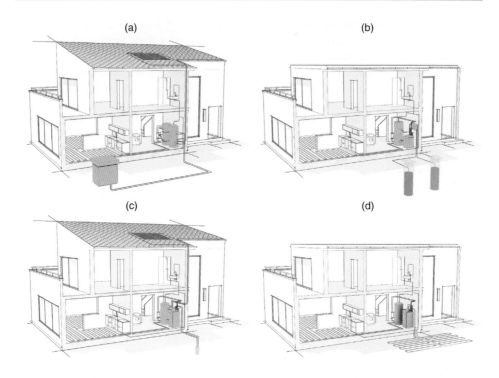

Fig. 3.4 Schematic drawing of (a) air source heat pump, (b) groundwater heat pump, and (c) ground source heat pump with borehole and (d) with horizontal ground heat exchanger. (Courtesy of Stiebel Eltron AG)

heat pump itself as shown in Figure 3.4c and d. The average annual performance[1]) measured in recent field studies for air source heat pumps in Central Europe is about $SPF_{HP} = 2.9$, whereas an average $SPF_{HP} = 3.9$ has been measured for borehole heat sources and $SPF_{HP} = 3.7$ for horizontal ground heat exchangers [22,23]. These differences are due to higher temperatures of the ground heat source at the times of the year where most heat is demanded. Losses that occur for the defrosting of the air-to-refrigerant heat exchanger may be another reason for the lower SPF of the air source units. The deviation of the performance of a particular installation from the average of field test measurements may be quite large. Partially, this can be due to differences in the performance of the heat pumps on the market, but to an even larger degree this may be caused by higher or lower temperatures of the heat source and of the useful heat that has to be delivered. As a general rule of thumb, a decrease of the temperature difference between the evaporation and the condensation by 1 K leads to about 2–3% increase of the COP under typical operating conditions.

Recent developments in heat pump technologies, without going into details, include

– the use of economizer cycles to overcome larger temperature differences between the source and the sink;

---

[1]) For the definition of seasonal performance factor (SPF), see Chapter 4.

- the use of desuperheaters to make use of higher exergy heat for DHW preparation while the condenser is operated at the temperatures of a floor heating system;
- the use of capacity-controlled compressors, especially for air source heat pumps, to decrease the mismatch between the heating power and the required power;
- the phase-out of refrigerants with significant ozone depletion potential and the development of refrigerants with lower global warming potential;
- the development of low temperature lift compressors with high COPs, for example, turbocompressors [24];
- three-fluid evaporators for direct heat transfer of two different heat sources with the refrigerant [25].

Simulation models for heat pumps have been reviewed for T44A38 by Dott *et al.* [26]. A review on heat pump and chiller models has also been given by Jin and Spitler [27]. In standards, mostly easy-to-use calculation methods are required for the SPFs of commonly used heat pumps. For the evaluation of new and more sophisticated system concepts, a more detailed modeling is required to be able to consider system dynamics or to evaluate the systems under varying boundary conditions. Therein, the interaction of heat loads such as building or domestic hot water demand with heat storages and heat sources, for example, borehole heat exchangers or solar heat, plays a key role for the evaluation of the system behavior over long-term periods such as whole years or short-term periods to evaluate, for example, the control strategies.

Empirical black box models are quite widespread. The reason for this is that the representation of the component behavior in the system is sufficiently precise and that the required data of individual products are mostly available. Physical models, or models based on physical effects, are not widely used for annual energy performance simulation since the required computation time rises significantly for solving the states and flows of the refrigerant cycle for each simulation time step. Therefore, quasi-steady-state performance map models (i.e., black box models) are the most widespread heat pump models in dynamic simulation programs such as TRNSYS, ESP-r, Insel, EnergyPlus, IDA ICE, or MATLAB/Simulink Blocksets. Examples for these models can be found in Ref. [28] and are also implemented in the simulation software Polysun [29]. Therein, a restricted number of sampling points from performance map measurements are used either to interpolate in-between those points or to fit a two-dimensional polynomial plane (see Figure 3.5a). These models use the inlet temperature of the heat source and the desired outlet temperature on the heat sink side of the heat pump to calculate the thermal output and the electricity demand. The extension of black box steady-state models for the inclusion of dynamic effects such as for icing/defrosting and for the thermal inertia in the condenser or evaporator has been described, for example, in Ref. [28].

More complex models are available that calculate the performance of the heat pump based on the performance of the compressor and the overall heat transfer coefficients of the evaporator and the condenser [27,30–34]. The compressor may thereby be simulated based on assumptions for the volumetric and isentropic efficiency or based on a performance map that can be obtained from the manufacturer of the compressor. These models have the advantage that they are more flexible and can thus be used to study changes in the heat pump refrigerant cycle such as the inclusion of two evaporator heat

3.2 Heat pumps

(a)

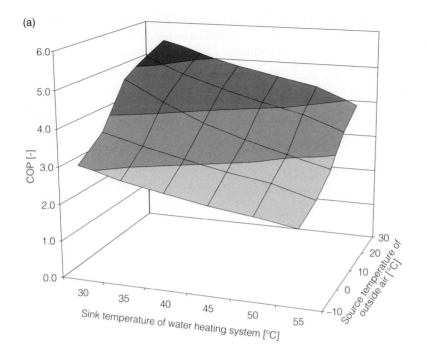

(b)

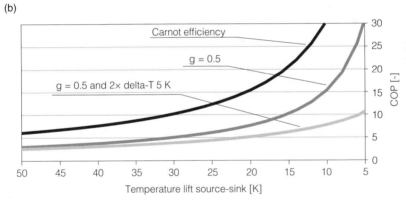

Fig. 3.5 (a) Exemplary COP performance map of an air-to-water heat pump. (*Source*: [26].) (b) Calculated COP of a heat pump based on source and sink temperature lift, the Carnot efficiency, and exergetic efficiency *g*, with and without adding a temperature difference (delta-*T*) from source and sink to the refrigerant

exchangers in series – one for the use of air as a heat source and the second one for the use of brine from a solar heat source, and/or an additional desuperheater to provide DHW while the heat pump delivers space heat. These additional model features may justify the higher computational effort that is needed to compute the thermodynamic states of the refrigerant in the heat pump cycle iteratively.

For the modeling of heat pumps that can take heat from solar collectors for the evaporator and may thus run on higher source temperature levels than usual, special attention has to be paid in order not to overestimate the performance of the heat pump in this application:

- A very simple approach for a black box model is to assume that the COP of the entire heat pump is a constant fraction of the thermodynamically maximum Carnot efficiency. However, an extrapolation of COP values with this approach to temperature lifts that are much lower or much higher than the ones for which this model is calibrated cannot be expected to produce reliable results. Furthermore, if the temperature lift is taken as the difference between the temperatures of the heat source and the heat sink, it has to be taken into account that this is not equal to the temperature lift between the evaporation and the condensation of the refrigerant. The temperature difference within the respective heat exchangers has to be taken into account. Thus, an extrapolation of the heat pump performance to low temperature lift operation using a constant fraction of the Carnot efficiency is likely to overestimate the heat pump performance quite significantly [35] (Figure 3.5b).
- Some heat pump models that calculate the state of the refrigerant at different points in the cycle are based on the assumption that the temperature difference for the superheating after the evaporation of the refrigerant is a constant value. However, especially for heat pumps that are equipped with thermostatic valves, the amount of superheating after the evaporation process may increase with increasing source temperatures and consequently reduce the achieved COP under these operating conditions [34].

Simulation models for capacity-controlled heat pumps have been presented in Refs [33,36–40]. However, only little data are available for the validation of capacity-controlled heat pump models (Table 3.2). In general, the lack of data availability for the parameterization of physical models and especially also for modeling capacity-controlled heat pumps is currently limiting the use of these models for heat pump system development projects. The same lack of data is observed for heat pump operation with unconventionally high source temperatures.

3.3 Ground heat exchangers

Ground-coupled heat pumps use the ground, groundwater, and surface water as a heat source, or in case of cooling as a heat sink.

Ground-coupled heat pump systems are increasingly used worldwide due to their high potential of energy savings and CO_2 emission reduction. The reason of the better performance of ground-coupled heat pumps compared with air source heat pumps can be explained from the ground characteristics. The undisturbed ground temperature as a function of depth, derived from the analytical equation of the heat conduction, is shown in Figure 3.6. The undisturbed temperature in the ground from 6 to 50 or 100 m depth is usually close to the mean annual surface temperature at the specific location. Therefore, it is warmer than the ambient air in winter, cooler in summer, and much more constant over the year, thus providing a better source of energy for the heat pump than air. This is especially true on days of extreme climatic conditions when the heating or cooling demand is highest. The performance of ground-coupled heat

3.3 Ground heat exchangers

Table 3.2 Heat pump models (nonexhaustive)

Name and ID	HP type	Model type	Transient effects	Capacity control	Comments	References
Simulation models						
TRNSYS Type 877	All	Ref-Cycle, C-Perf or (η_V and η_S), ε-NTU	QSTAT, Cond-Cap	√	a,b	[34]
RDmes Carbonell	W/W, B/W	Ref-Cycle, η_V and η_S, ε-NTU	QSTAT	—	c	[27,41]
TRNSYS Type 176	All	Ref-Cycle, C-Perf, ε-NTU		√		[30]
TRNSYS Type 372		Ref-Cycle, C-Perf	QSTAT	—		[42]
Madani EES – model		Ref-Cycle, C-Perf	QSTAT	√	b,d	[33]
TRNSYS Type 401 (201) (YUM)	All	HP-Perf (biquadratic fit)	Evap-Cap, Cond-Cap	—		[43]
TRNSYS Type 204 (YUM)	All	HP-Perf (biquadratic curve fit)	QSTAT, Cond-Cap	—	e	[44]
TRNSYS Types 504, 505, 665, 668	All	HP-Perf		—		[45]
Carnot HP	All	HP-Perf	Evap-Cap, Cond-Cap	—		[46]
Polysun	All	HP-Perf (interpolation)		—		[19]
Temperature bin calculation						
EN 15316-4-2:2008	All	HP-Perf		—		EN 15316-4-2:2008
EN 14825:2012	All	HP-Perf		√		EN 14825:2012
ANSI/ASHRAE Standard 137-2009	All	HP-Perf		—		ANSI/ASHRAE Standard 137-2009

Ref-Cycle = simulation of thermodynamic state of the refrigerant at different points in the cycle; C-Perf = compressor performance map (interpolation from single points or curve fits); HP-Perf = heat pump performance map (interpolation from single points or curve fits); QSTAT = quasi-static; Cond-Cap = heat capacitance of the condenser; ε-NTU = effectiveness–NTU model for heat exchangers.
a) Defrosting losses implemented as static reduction of heat output dependent on RH and deicing efficiency.
b) Including desuperheater simulation.
c) Model parameters identified automatically based on catalog data.
d) Separate models for variable-speed compressor, single-speed compressor, evaporator, and condenser.
e) Dual-stage compressor (two performance maps) heat pump including defrosting losses.

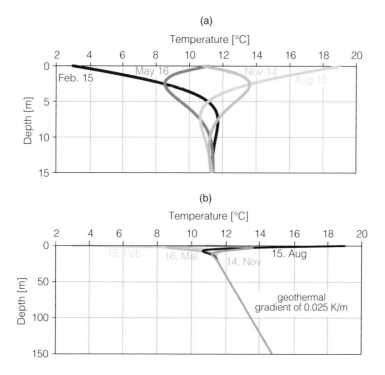

Fig. 3.6 Temperature of the undisturbed ground for a selected northern hemisphere climate as a function of the depth and time of the year for shallow depths (a) and large depths (b)

pumps is about 20–30% higher than that of equivalent air source heat pumps. A better performance on the overall system level of both types of systems can be achieved by combining them with solar thermal technology that reduces the operating hours and thus the electricity demand of the heat pump.

Ground source heat pumps can have open source loops using air as a heat carrier or a water reservoir as a direct energy source. Closed loops can be classified as indirect systems, if a ground heat exchanger (GHX) is linked to a water or brine source heat pump, or direct expansion systems, if the refrigerant of the heat pump circulates directly through the ground heat exchanger. Direct expansion systems are more efficient but also more difficult to install and therefore indirect systems are used more frequently. Closed loop systems are also referred to as ground-coupled heat pumps and are the main focus of this section.

Compared with air source heat exchangers, GHXs do not have to be defrosted and the heat pump compressor suffers from less mechanical and thermal stress due to more stable source temperatures. Thus, the reliability is higher and a lifetime of about 20–25 years can be expected for ground-coupled heat pumps. The ground coils, typically made of polyethylene or polybutylene, have a life expectancy of 50 years and more [47].

3.3 Ground heat exchangers

GHXs can be classified into horizontal and vertical (HGHXs and VGHXs) as a function of their physical extension that determines also the depth at which the heat exchange takes place. Examples of both configurations are presented in Figure 3.4. Usually, VGHXs are more efficient. They need less land area and less pumping energy due to the better thermal characteristics of the ground at large depths. Therefore, they are more suitable for large systems. However, the HGHXs are usually cheaper to install and bear less risk during construction than borehole drilling. The heat extracted from the ground is usually regenerated slowly from the earth's surface rather than from deeper geothermal sources. Exceptions of this rule are possible for extreme depths or for special geological conditions.

VGHXs consist of vertically drilled boreholes at typical length range between 45 and 150 m with diameters of 10–15 cm. Each borehole is equipped with concentric, U-tube, or double U-tube thermoplastic pipes with diameters of 2–4 cm and backfilled with a grout material that enhances the heat transfer and is usually made of sand and bentonite. In order to limit thermal interference, the distance between boreholes should at least be 5–7 m, depending on ground characteristics. Higher heat pump source temperature can be supplied by increasing the depth of boreholes due to increasing natural ground temperature. Construction of very deep boreholes is the topic of current research [48]. Insulation of the upper part of a deep borehole has been proposed in order to achieve higher temperatures of extraction [49]. In the case of multiple boreholes or borehole fields, a ground storage volume can be delimited that may require regeneration over each year.

HGHXs are characterized for being placed at shallow depths, usually between 1 and 3 m. The HGHXs can be classified into three groups: (i) horizontal, (ii) vertical oriented, and (iii) building integrated. Some examples of horizontal orientations are meander, harp, bifilar, capillary pipes, and earth to air heat exchangers. Trench, cage, and basket/helix are examples of vertical oriented types. Moreover, the HGHXs can be integrated in the basement of a building, in the walls, or as energy piles. Illustrative examples of some of these configurations can be found in Ref. [50]. For building integrated systems, thermal short-circuiting must be considered and may reduce the performance of the system significantly.

The initial capital cost of a ground source heat pump system compared with an air source heat pump system is around 30–40% higher due to the cost of drilling the borehole and installing the ground coil. Historically, the length of the borehole has been oversized, thus increasing the initial capital and operating costs. It is thought that this initial cost is an important barrier for this technology to develop faster. Therefore, there is a need to have robust simulation tools able to size and predict the behavior of the GHX with accuracy at acceptable computation time. It is also very important to know the ground properties and groundwater conditions well; otherwise, a simulation does not deliver reliable results.

The sizing of GHX is usually done such that the fluid inlet always remains above a defined temperature limit. Both the energy of net annual heat extraction and the nominal power of heat extraction influence this temperature. Furthermore, this temperature decreases over months (HGHX), over years (VGHX), and even over decades in the case of a field of VGHXs. Therefore, the calculated minimum inlet temperature after several years is typically used to determine the size of the GHX. Assuming that building loads are known, one of the main uncertainties in the modeling of GHX is the knowledge of the ground properties and possibly also of natural groundwater flows. The ground conductivity and borehole effective thermal resistance have a large influence on the performance of the GHX. The ground properties either are derived from geologic maps, often provided by local authorities, or can be measured with a geothermal response test.

One fundamental difference between modeling VGHXs and HGHXs is the influence of seasonal temperature variations at the ground surface. These affect considerably the performance of HGHXs but can be neglected for most VGHXs (Figure 3.6). Consequently, the simplifying assumptions for modeling the GHX differ quite substantially between VGHXs and HGHXs. A detailed review of GHX models can be found in a technical report of T44A38 [50]. Hereafter, a short overview is provided.

3.3.1 Modeling of vertical ground heat exchangers

A common approach for modeling VGHXs consists in splitting the calculation into a far-field problem (also called global problem) and a near-field problem (also called local problem) (Table 3.3). The near field, the borehole and its close surroundings, is affected by short-term changes in heat extraction or injection as well as by heat transfer between the upward and downward flowing fluid and may be solved on a small time step basis from minutes to hours. The far-field problem determines the temperature at the outer boundary of the near field after a certain amount of time (time steps of days or even months). It depends on axial effects and thermal interference between multiple boreholes.

In order to solve the near-field problem of a GHX, it is necessary to use a submodel for calculating the thermal resistance of the borehole that depends on the grout properties, pipe material, circulating fluid, and geometry of the pipe legs. A detailed review of the most important models for the borehole resistance can be found in Ref. [67].

Popular solutions for the far-field problem are g-functions proposed by Eskilson [68] that were derived using numerical simulations. These functions are defined as the nondimensional thermal response factors of a borehole field with constant heat extraction/injection. Once a g-function is known, the temperature difference between the outer margin of the borehole and the undisturbed ground after a time of known constant heat extraction or injection can be calculated based on the thermal properties of the ground. Several boreholes in a field can be solved by spatial superposition and calculations with variable heat extraction/injection can be conducted by means of temporal superposition. In order to predict transient effects that are shorter than a couple of hours, so-called short-time g-functions were presented by Yavuzturk and Spitler [69].

For modeling VGHXs, the easiest and efficient approach in terms of computational time consists in using analytical models. Among them, the most accurate solution is the finite line source model proposed by Claesson and Eskilson [70] and numerically solved by

3.3 Ground heat exchangers

Table 3.3 Models for vertical ground heat exchangers (nonexhaustive)

Platform and ID	Use	Coax	U	2U	Fields	Far-f	Near-f	Bore-Cap	Grd-Prop	GWF	Documentation (D)/ validation (V)
TRNSYS Type 557 (DST)	AS	√	√	√	√	3D FD	1D FD	—	N	—	D: [51]; V: [52]
TRNSYS Type 451	AS	—	—	√	—	ILS	1D FD	√	N	—	[53,54]
TRNSYS DSTP Type 280	AS	√	√	√	√	3D FD	1D FD	—	N	$\surd^a$	D: [51,55]; V: [56]
TRNSYS SBM Type 281	AS	√	√	√	√	3D FD	1D FD	—	1	—	D: [57,58]; V: [59]
PILESIM2 (stand-alone)	Size	√	√	√	√	3D FD	1D FD	?	N	$\surd^a$	D: [60]; V: [61]
EED (stand-alone)	Size	√	√	√	√	g (FLS)	Anal	—	1	—	[62]
GLHEPRO (stand-alone)	Size	√		√	√	g (FLS)	Anal	—	1	—	[63]
EWS (stand-alone)	Size	√		√	√	g (FLS)	1D FD	√	N	—	[64]
Polysun EWS	AS	√	√	√	√	g (FLS)	1D FD	√	10	—	[64]
Simulink EWS	AS	√		√	√	g (FLS)	1D FD	√	N	—	[65,66]

AS = annual simulation; size = sizing of boreholes; coax = coaxial tube; U = U-tube; 2U = double U-tube; fields = field of several boreholes influencing each other; far-f: approach for far-field temperature development; near-f: approach for near-field temperature development; Grd-Prop = number of horizontal layers for which ground properties can be defined; GWF = influence of groundwater flow; 1D/3D = one-dimensional/three-dimensional; FD = finite difference; FLS = finite line source; ILS = infinite line source.
$a)$ Influence of groundwater flow computed using important simplifications – not validated.

Lamarche and Beauchamp [71]. The current finite line source models, including also inclined boreholes, and are used to calculate g-functions with similar accuracy as the numerical models; therefore, analytical models may be recommended for long-term analysis. Most of the analytical models presented in the literature are not suited for short-term analysis predictions and, only recently, some studies have addressed this issue [71,72].

The main difference between deriving the g-functions from numerical or analytical models is that for numerical models the g-functions must be preprocessed, while the analytical g-functions can be calculated on the fly during each simulation. Nevertheless, numerical models are more often used for short-term analysis and the development and use of analytical models for short term is still a topic of current investigations.

3.3.2 Modeling of horizontal ground heat exchangers

Due to the shallow depth, the HGHXs are strongly influenced by weather conditions such as variation of the ambient temperature, thermal radiation (solar and long wave), rain, and snow. In addition, freezing of water in the vicinity of the pipes may play an important role in improving the GHX performance due to the latent heat release, thermal storage, and increase of ground thermal conductivity of the frozen region [73]. Ice formation is of relevance for most HGHXs placed in moist ground when the return temperatures to the ground may be below 0 °C for several weeks in the year. Different models for the annual performance simulation of heating systems have been described in the literature (Table 3.4). These models are generally divided into two groups mainly depending on the fluid carrier and type of loop: (i) open loops with earth to air heat exchangers and (ii) closed loops with water/glycol using shallow ground heat exchangers. The earth to air heat exchangers employed in the building sector are mostly used for supply air preheating or cooling.

Most of the models to solve HGHXs with closed loops before the 1990s were based on the line source theory with a consequence of oversizing the GHX length. More elaborated models are based on the solution of the heat conduction equation of one single pipe in cylindrical coordinates [73,74] or of a pipe array in two-dimensional Cartesian coordinates in a plane that is perpendicular to the fluid flow [75,76]. For most configurations, instead of using bidimensional numerical approaches, the HGHXs can

Table 3.4 Models for horizontal ground heat exchangers (nonexhaustive)

Platform and ID	HX type	Model type	Grd-Prop	Special effects				Comments	Documentation (D)/ validation (V)
				Ice	Ihor	Rain	GWF		
TRNSYS Type 556/ORNL	H	3D FD	1	—	—	—	—		D/V: [73]; D: [74]
Glück (stand-alone)	H	2D FD	N	√	√	—	—		D: [75]
TRNSYS – Ramming	H	2D FD?	N	√	√	√	—		D: [76]
TRNSYS Type 460	H	3D FD?	3D	—	—	—	—	a	D/V: [77]
Simulink 1D	H	1D FD	N	√	√	—	—		D/V: [78]
Simulink 2D	H, B, T	2D FEM	2D	√	√	—	—		D: [79]

H = horizontal pipes; B = basket; T = trench; ice = influence of ice formation; Ihor = time-dependent radiative surface gains; rain = influence of rain; GWF = influence of groundwater flow.
a) Hypocaust/sensible and latent heat transfer of moist air in an earth duct.

be modeled combining analytical models for the fluid flow with a one-dimensional model for the ground [65].

3.3.3 Combining GHX with solar collectors

The combination of GHX with solar thermal collectors offers the possibility to use the solar energy directly, as a source for the heat pump, or for regenerating the ground. Ground regeneration by solar energy offers the advantage to increase the temperature level of the ground or to suppress the long-term degradation effect for imbalanced heating/cooling loads that increases with the number of boreholes in the vicinity. During the last few years, several studies have analyzed the possible benefits of regenerating boreholes. In principle, it has been stated that direct use of solar energy, at least in summer, provides better SPF compared with systems recharging the ground. The benefit of regeneration is low and eventually not worth the effort for single boreholes that do not suffer dramatically from degradation of the heat source with time (i.e., the maximum degradation is in the range of 1–3 K), but it may be required for continuous long-term operation of larger borehole fields where degradation of the heat source with time continues even after 10 years and may be substantially more than 10 K ([80], p. 34ff). It is emphasized that multiple single and adjacent GHXs can also be seen as a GHX field, even if they are commonly not calculated as such. Also systems with a few, but undersized (i.e., too short) GHXs have been found to benefit from ground regeneration [81]. In these cases, attention must be paid to the electric consumption of the GHX pump and low flow rates are recommended for regeneration [82]. For ground that is largely influenced by groundwater flow, regeneration may be obsolete. Furthermore, it is often claimed that the temperatures for recharging the ground have to be kept below a certain limit to avoid drying out of the ground, stress on pipe material, groundwater quality concerns, or other problems. However, the authors were not able to find detailed information or scientific evidence for such concerns.

3.4 Storage

In order to overcome mismatch of heat production and heat consumption, solar and heat pump systems are often equipped with thermal energy storage. The most frequently used storage devices are based on sensible heat, mostly using water as a storage medium. Furthermore, phase change processes (solid–liquid and liquid–gas) and thermochemical processes based on heat of sorption and desorption may also be used for storing heat. Expectations to be able to increase the storage density and/or efficiency with phase change materials (PCMs) or with sorption processes have led to increased research in these two fields. However, the only phase change storage technology that has achieved a solid market presence with validated economic benefits for HVAC systems up to now is the water–ice phase change storage. Sorption processes are used in thermally driven heat pumps that make use of waste heat for cooling and may also be used for solar thermal cooling. The use of sorption processes for seasonal heat storage is investigated in research projects [83–85], but no commercial solutions are yet available. Thermochemical storage can also be achieved by reversible chemical reactions. The energy density may be up to 10 times more than that for sensible heat storage, but no proven technology has yet been introduced into the market for the range of 100–150 °C, which is a reasonable maximum temperature range for solar installations in dwellings. Research efforts in this field have been undertaken by the IEA SHC Task 32 that was later followed by the IEA SHC Task 42 [86,87].

3.4.1 Sensible heat storage and storage in general

Sensible heat storage may be based on solid or on liquid material. Sensible storage in liquids includes the well-known water storage tanks as well as salt gradient solar ponds [88,89], but also synthetic oils or molten salts have been used for higher temperatures [88]. Solid materials include rock bed thermal storage [90] and concrete, which includes also the thermal activation of building elements such as concrete floors. Storage solutions that include solids as well as liquids are aquifer thermal energy storage, cavern thermal energy storage, gravel–water thermal energy storage, and borehole thermal energy storage [91,92]. Storage in the ground is only treated in this chapter for the case of ground buried storage containers, whereas storage in the ground using boreholes is a topic treated partially in Section 3.3.

Most solar and heat pump applications use water as a storage medium. Water is an excellent sensible heat storage medium due its large specific heat, low price, nontoxicity, and chemical inertia. The specific heat of water (around 4.19 kJ/(kg K)) is still unbeaten. However, compared with the storage of energy in fuels such as hydrocarbons or hydrogen, the amount of heat that can be stored in water is low and the exergetic value is rather poor. In the temperature range of 20–80 °C that is suitable for domestic applications, the sensible heat storage capacity of water is 70 kWh/m^3, whereas the energy contained in heating oil is roughly 100 times more and the exergetic value of heating oil is around 500 times more. Because of this comparatively low storage density and the additional aspect of heat losses that increase with time, the use of water for storing sensible heat is economically attractive for short-term storage, but less so for seasonal storage.

DHW storages are used not only for demand peak shaving, but also for storage of solar heat that is produced at times where there is no demand. Storage may also be used for space heating to prolong the running times of heat pumps when the space heat load is below the heating power of the heat pump, and also for storing solar heat. So-called combi-storages can be used for storing DHW and space heat in one single unit. The advantage of this solution is a better surface to volume ratio compared with a two-storage solution and only one device where the solar heat has to be conveyed, which simplifies hydraulics and control. Different concepts for preparing DHW from a combi-storage are shown in Figure 3.7. The combi-storage concept relies on the temperature stratification of water that is a result of its temperature-dependent density. Thus, hot water at the top is used for delivering hot DHW to the consumer, low temperatures at the bottom are used for solar preheating of cold water from the mains, and the region in the middle is available for space heating. Internal or external heat exchangers are used to transfer the solar heat from the antifreeze fluid in the collector loop to the storage medium and to transfer heat from the storage water to the DHW distribution system.

3.4 Storage

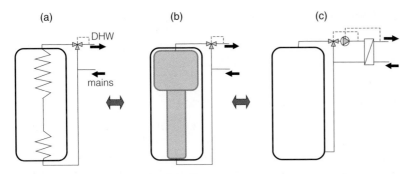

Fig. 3.7 Different variants for DHW preparation from a solar combi-storage: (a) immersed HX spiral; (b) tank-in-tank; and (c) external DHW module. (*Source*: Ref. [93])

Water storage vessels are most often pressurized steel vessels, but can be designed unpressurized or from other materials such as aluminum or polymers, with possible cost benefits. However, unpressurized vessels might require additional heat exchangers to transfer heat into pressurized systems and may be more prone to corrosion, particularly if they are open to the atmosphere. The lifetime of pressurized storage vessels is typically 20–80 years, but can be much shorter if design errors lead to accelerated corrosion. Key issues that have been identified for efficient storage solutions are

- good insulation (typical U-values in the range of maximum 0.3–0.5 W/(m² K)) without thermal bridges;
- non-return valves or other means to prevent unwanted buoyancy-driven circulation in connected loops (e.g., collector field at night, space heating loop in summer, or auxiliary heater loops);
- heat traps on all storage connections to avoid the so-called counterflow heat losses in standby pipes without net volume flow.

The capability to achieve and maintain stratification within the storage is particularly crucial for combi-storages, and of utmost importance if a heat pump is used to charge this storage.

Charging and discharging of storages is achieved by circulating a heat carrier fluid directly through the storage (direct charging/discharging) or through an internal heat exchanger in the storage. Charging may additionally be achieved with internal heat sources such as immersed electrical heaters or integrated fuel burners. In accordance with Figure 3.8, Equation 3.6 presents the general form of the energy balance equation for thermal energy storage:

$$\dot{Q}_{HX} + \dot{Q}_{port} + \dot{Q}_{gen} - \dot{Q}_{loss} = m \frac{dh}{d\tau}, \tag{3.6}$$

where $\dot{Q}_{HX}$ and $\dot{Q}_{port}$ are the net heat transfer rate to the storage of all internal HXs and direct inlet/outlet port(s), respectively, $\dot{Q}_{gen}$ is heat generated inside the storage, and

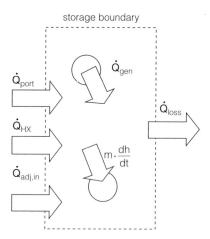

Fig. 3.8 The energy balance for a thermal storage

$\dot{Q}_{\text{loss}}$ is the heat loss from the thermal storage to the ambient. The change rate of the stored energy is the product of the storage mass m and the change rate of the specific enthalpy $dh/d\tau$. For the modeling of stratified thermal energy storage, the thermal mass is divided into a series of nodes with proper spatial arrangement for finite differential analysis. For each node, the energy balance must also account for the exchange of heat with adjacent nodes as shown in Equation 3.7 and Figure 3.8:

$$\dot{Q}_{\text{HX},i} + \dot{Q}_{\text{port},i} + \dot{Q}_{\text{adj,in},i} + \dot{Q}_{\text{gen},i} - \dot{Q}_{\text{loss},i} = m_i \frac{dh_i}{d\tau}, \qquad (3.7)$$

where $\dot{Q}_{\text{adj,in},i}$ represents the net heat gain from all adjacent storage nodes. This net heat gain accounts for mass transfer as well as diffusion and thermal conduction processes. In the case of PCM modules included in a storage, $\dot{Q}_{\text{adj,in},i}$ is also used in the so-called enthalpy method to account also for heat transfer from and to the PCM modules. As long as the specific enthalpy–temperature curve is obtained for certain PCM, the condition of each node can be solved.

Short-term behavior and particularly effects of mixing and stratification during charging and discharging of water storages can be simulated based on 3D finite volume methods with computational fluid dynamics (CFD) [94–98]. The computational effort for such simulations is large with computation times of several hours or even days for the simulation of 1 h in the life of a storage vessel. For annual performance prediction, one-dimensional methods (finite difference lumped parameter models) are usually used that require only a few seconds or minutes for the simulation of a whole year, but with limited accuracy. However, one-dimensional models usually make important simplifications such as homogenous temperatures in the horizontal direction and no consideration of mixing effects caused by turbulent inflows.

Two different approaches may be distinguished for one-dimensional models: the fixed node approach uses volume elements that are fixed in the space domain and usually

3.4 Storage

of the same volume each. Thus, when fluid moves along the vertical axis of the stratifying storage tank model it is transferred from one volume element to the other, thereby leading to "numerical diffusion" if the Courant number is not equal to 1. Numerical diffusion can be avoided to a great extent using plug flow models where each charging mass flow produces a new fluid plug that is exactly of the size that corresponds to the simulation time step multiplied by the volume flow rate at the inlet. These fluid plugs move along the vertical axis of the storage as charging and discharging occurs.

Numerical diffusion of fixed node simulation approaches is to some extent counterbalanced by not taking into account other diffusion processes in the storage [99], and the plug flow approach may overpredict the stratification capability of the storage if these other diffusion processes are not included in the model. Although the inclusion of effects of inlet jet mixing or plume entrainment in one-dimensional models has been presented for some cases [96,100,101], these approaches are still lacking validation for a larger set of boundary conditions. Thus, the options for influencing the stratification efficiency of one-dimensional storage models shown in Table 3.3 are usually restricted to (a) reducing the number of fixed volume elements (nodes), thereby increasing numerical diffusion, (b) iteratively mixing the entering fluid with adjacent fluid volume elements in case of temperature inversion, or modeling ideal stratification by adding the inflowing fluid element to the storage element with the least deviation in temperature, and (c) adjusting the effective thermal diffusivity of the storage medium empirically.

Other important features that distinguish implemented storage models are the number of heat exchangers that can be simulated, the number of direct storage charging connections ("double ports") that can be simulated, and the ability to simulate tank-in-tank or mantle tank combi-storages.

Some models assume a purely temperature-dependent stratification [96] instead of using the temperature-dependent density. These models are are not able to show the mixing and inversion of the temperature profile that occurs if water is cooled below 4 °C.

3.4.2 Latent storage

The most frequently used latent storage technique is the phase transition of water to ice and vice versa, which has been in use for a long time for storage of cold (solid = charged) and has recently been introduced for storage of heat (liquid = charged) in solar and heat pump applications where solar heat is used for charging and the heat pump is used for discharging. Ice storages for solar and heat pump systems may be small vessels (around 300 l) inside the building [102] or large (several m^3) and buried in the ground outside of the building [103]. The formation of ice on the heat transfer surfaces leads to increased heat transfer resistance with time due to the increasing ice layer. Therefore, large heat transfer surfaces must be used or alternatively the ice is removed from time to time [104].

Other phase change materials with higher melting temperatures may be used on the hot side of the heat pump. A general overview of PCM for energy storage can be found in

Refs [105,106]. The used materials typically have a low thermal conductivity, which puts limits to possible heat transfer rates. This problem can be overcome by adding material with higher conductivity, for example, graphite, and by increasing the ratio of heat transfer surface to volume by

- macroencapsulation into containers, which can have different shapes, for example, balls or cylinders [107];
- microencapsulation into capsules of only 5–10 µm; the microcapsules can be mixed with sensible heat transfer fluid (such as water) to form PCM slurries that can be pumped [108]; or
- multilayer PCM units and bulk PCM tanks with integrated fin-and-tube heat exchangers [109–111].

Quite often, hybrid PCM and sensible storage units are used. These are, for example, water tanks with immersed macroencapsulated PCMs [112,113].

3.4.3 Thermochemical reactions and sorption storage

Heat can be absorbed or released by chemical compounds when they interact under certain conditions. The dissociation of a compound into its constitutive components is usually endothermic, and the recombination is exothermic. Thus, in the charging process the absorbed energy is used for dissociation of the material, and is equivalent to the heat of reaction or the enthalpy of formation. This principle can be used to store heat, for example, during sunny periods and release it when needed.

During the chemical storage reaction, described as $C + heat = A + B$, C is called the thermochemical material for the reaction, and materials A and B are called the reactants. A is usually a hydrate, hydroxide, carbonate, ammoniate, and so on and B can be ammonia, water, carbon monoxide, hydrogen, and so on. Most often C is solid or liquid, while A and B can be any phase.

Several reactions are candidate for this process, but many have been studied in detail for the purpose of storing solar heat at rather low temperature (20–200 °C). No commercially available solution is yet available since main criteria for a sustainable solution, such as reversibility, nontoxicity, low cost, and limited volumes, are not yet fulfilled.

Physical sorption reactions may also release heat. In the sorption process, a sorbent absorbs (liquid sorbent) or adsorbs (solid sorbent) a sorbate. Desorption refers to the separation of the sorbate from the sorbent, which requires the supply of (usually higher temperature) heat. Sorption processes are often referred to as chemical heat pump processes since in order to be available to release useful heat they need a lower exergy heat source [114]. In open processes, this is not always apparent at first sight since the lower exergy heat may be hidden, for example, in water vapor that crosses the process boundaries. Theoretically, sorption storage has the potential of increasing the storage density by a factor of 3–4 compared with the sensible heat low-pressure water storage tank. At the same time, once the storage is charged (sorbent and sorbate are separated) they do not suffer from further losses of energy or exergy. This seems to be attractive especially for long-term heat storage and many research projects are

3.5 Special aspects of combined solar and heat pump systems

Table 3.5 Storage models for water tanks

Name and ID	Nodes (N) or plugs (P)	HX	DP	Tank-in-tank	Mantle tank	Internal electric heaters	Stratification options	Heating or cooling	Comments	Documentation (D)/validation (V)
TRN Type 4	$N \leq 100$	0	2			2	PE, S	H, C	a,b	D: [119]
TRN Type 60	$N \leq 100$	3	2			2	PE, S	H, C	a,b	D: [119]
TRN Type 38	P var.	0	2			1	PE, S	H, C	a,b	D: [119]
TRN Type 340	$N \leq 200$	4	10	✓	✓	1	PE, S	H	a,b	D: [120]
Polysun storage	$N = 12$	6	5	✓	✓	3	IJM, PE, S	H	a,b	D: [19]

C = cooling; H = heating; PE = plume entrainment; IJM = inlet jet mixing; S = stratified; TRN = TRNSYS.
a) Domestic use.
b) Solar energy.

ongoing on this topic. However, this technology has reached commercial stage only for heat-driven heat pumps, but not yet for long-term storage, and is thus not further elucidated in this book.

Examples for simulation models for different types of thermal storage are given in Tables 3.5 and 3.6.

3.5 Special aspects of combined solar and heat pump systems

3.5.1 Parallel versus series collector heat use

A heat pump may use heat from a solar collector as an energy source for the evaporator. This mode of operation is referred to as a series configuration [20] or as indirect collector heat use. With the series configuration, the collector may be operated at or below the ambient air temperature. In this case, instead of losing heat to the ambient air the collector may even gain heat from the ambient air. In general, the solar collector yield is always higher when it is operated at lower temperatures. At the same time, the COP of the heat pump rises with the temperature of the heat source. Therefore, the COP of the heat pump may be higher when solar heat from the collectors is used instead of ambient air.

In principle, solar heat can be used on both sides of the heat pumping process. Collector heat that is used in parallel concepts replaces directly heat supplied by the heat pump on the hot side of the heat pumping cycle. Collector heat on the cold side may lift the temperature level of the source and therefore improve the heat pump performance. In the absence of a heat pump, solar irradiance is lost if the collector temperature is not rising higher than the temperature required by the heat demand that is defined by the system of heat storage and heat distribution. This may be the case

Table 3.6 Storage models including phase change materials

Name and ID	Storage type	Nodes/ plugs/ plates	PCM shape					HX	DP	Heating or cooling	Comments	Documentation (D)/ validation (V)
			Cylinder	Sphere	Plate/ quader	Micro	Bulk + HX					
TRN Type 840	Water + PCM	N nodes	√	√	√	√		Total 5		H/C	a,b	D/V: [115]; V: [116]
Simard 2003	PCM	8 plates			√			1		C	c	D/V: [110]
TRN Type 841	PCM						√	1		H/C	a,b	D/V: [117]
TRN Type 60pcm	Water + PCM	N ≤ 100	√					3	2	H	a,b	D/V: [112]
TRN Type 60pcm2	Water + PCM	N ≤ 100	√	√	√			3	2	H	a,b	D/V: [113]
TRN Type 207	Water + ice									C		[118]

C = cooling; H = heating; TRN = TRNSYS.
a) Domestic use.
b) Solar energy.
c) Transport.

3.5 Special aspects of combined solar and heat pump systems

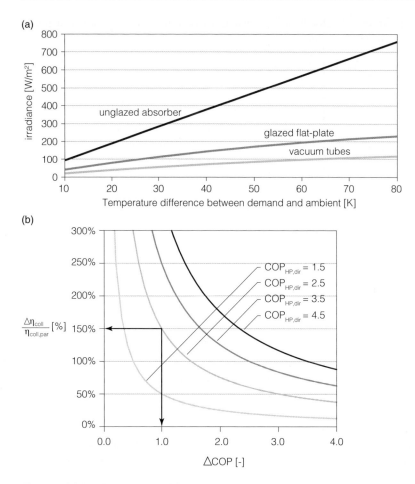

Fig. 3.9 (a) Irradiance needed for parallel collector heat use dependent on temperature difference between the heat demand and the ambient for three different collectors. (b) Curves for switching to series collector heat use for different values of $COP_{HP,par}$. (*Source*: Ref. [121])

when the temperature difference between the heat demand and the ambient is high while at the same time the irradiance on the collector field is low. This is illustrated in Figure 3.9a showing the irradiance on the collector field needed to make direct collector heat use possible.

The COP improvements that can be achieved by increasing the source temperature of the heat pump are from 0.5 to 2.5%/K and depend strongly on the heat pump operating point [122].

The combination of a higher COP and a higher collector yield is not yet a guarantee for a better overall system performance. Based on a mathematical analysis of a steady-state situation, Haller and Frank [121] found that only if the left-hand side of Equation 3.8 is greater than 1, indirect collector heat use instead of parallel collector and heat pump operation may improve the overall system's COP:

$$\frac{\Delta COP_{HP}}{(COP_{HP,par} - 1)} \cdot \frac{\Delta \eta_{coll}}{\eta_{coll,par}} > 1. \tag{3.8}$$

In this equation, $\eta_{coll,par}$ is the efficiency of the solar collectors if the heat is used in parallel to serve the demand. The increase in efficiency of the collectors that is achieved when the collector heat is used for the evaporator of the heat pump instead of parallel is $\Delta\eta_{coll} = \eta_{coll,ser} - \eta_{coll,par}$. Likewise, $COP_{HP,par}$ is the COP that would result if the heat pump would be operated parallel to the solar collectors and use a different source (e.g., ambient air). The increase in COP using the solar heat source is $\Delta COP_{HP} = COP_{HP,ser} - COP_{HP,par}$.

The implication of Equation 3.8 is illustrated in Figure 3.9b, showing curves that separate the area where series heat use is favorable (upper right) from the area where it is more advisable to use collector heat parallel. Assuming a COP of the heat pump using an air heat source ($COP_{HP,dir}$) of 2.5, an advantage from using collector heat for the evaporator of the heat pump is only possible if the COP of the heat pump increases by 1, and at the same time the collector efficiency increases by +150% relative to the direct collector heat use (See arrows in Figure 3.9b).

Obviously, increasing the collector yield by 150% (relative) implies that the collector yield is lower than 40% in parallel operation. Thus, the series operation is more advantageous for periods with low solar irradiance on the collector field. As a matter of fact, for each set of collector and heat pump efficiency parameters, there is a limit for the irradiance on the collector field above which series heat use would result in a lower SPF_{SHP}. This limit is dependent on the temperature of the heat demand and on the ambient air temperature [121]. However, it has to be kept in mind that these mathematical principles do not account for transient effects in the systems such as the influence of thermal capacities.

3.5.2 Exergetic efficiency and storage stratification

Low operating temperatures for solar thermal collectors or heat pump condensers increase the energetic performance of these components. Therefore, the efficiency of solar and heat pump systems decreases with increasing temperatures of the heat demand. Exergetic losses within the solar and heat pump system increase the temperature difference between the heat production and the heat supplied to the end use. Thus, these losses increase the operating temperatures of collectors and heat pumps and decrease the system efficiency. Examples for exergetic losses are the mixing of fluids with different temperatures on the one hand and heat transfer processes – for example, heat exchangers with large temperature differences between the primary and the secondary heat carrier fluid – on the other hand, as illustrated in Figure 3.10a.

Important sources for exergetic losses may be the storage tank as well as the hydraulics and control of the overall system. An ideal storage tank would be perfectly stratifying,

3.5 Special aspects of combined solar and heat pump systems

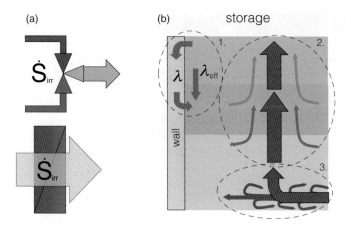

Fig. 3.10 Illustration of exergetic losses due to (a) hydraulics (mixing of two fluids with different temperatures) and heat exchangers and (b) processes in the storage tank including (1) thermal conduction and diffusion, (2) plume entrainment, and (3) inlet jet mixing

with no thermal diffusion. Three processes within the storage tank have been identified as the main drivers for reducing natural stratification [123]:

1. Thermal conduction and diffusion of water and other materials in the storage.
2. Plume entrainment caused by buoyancy-driven natural convection inside the storage where a thermal plume can be observed that entrains surrounding water.
3. Inlet jet mixing resulting from the kinetic energy of water entering the storage tank with high velocity.

Inlet jet mixing is of high concern especially for heat pumps that are combined with combi-storages. The COP of the heat pump benefits from high volume flow rates, but in turn the stratification efficiency of the storage may decrease with increasing flow rates due to inlet jet mixing. This is shown in Figure 3.11 for CFD simulations of a 2 in. diameter inlet where water is entering the storage tank with 30 °C. It can be seen that for a mass flow rate of 1800 kg/h (top) the negative effect on the higher temperatures in the upper half of the tank is dramatically more pronounced than for a mass flow rate of 360 kg/h. In other words, charging and discharging of the space heating zone causes a temperature drop at the level of the DHW temperature sensor in the DHW zone. As a result, the heat pump recharges the DHW zone of the storage with the corresponding higher temperatures, although only low-temperature heat would have been needed for space heating.

However, mixing processes occur not only in the storage, but also in the hydraulics of the system. For instance, hot fluid from the storage is mixed with cold fluid from the space heating return to reach the desired supply temperature for space heat distribution. For this reason, the parallel combination of heat distribution systems that include high-temperature radiators with low-temperature floor heating systems should be avoided.

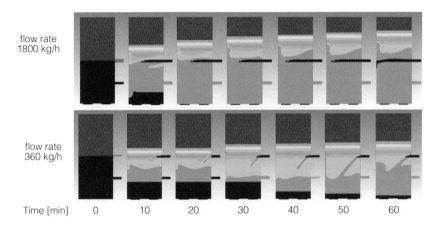

Fig. 3.11 CFD simulation results of an originally stratified storage (upper half: 50 °C; lower half: 30 °C) with an inlet (Ø 2″, 30 °C). (*Source*: Ref. [124])

In most solar combi-systems for domestic hot water production and space heating in the climate of Central Europe, 70–80% of the heat is supplied by the heat pump and only 20–30% is supplied by the solar thermal collectors. Therefore, the efficiency of the heat pump is crucial for a good overall performance. Exergetic losses of 10 K lead to 20–30% more electricity consumption of the heat pump. This illustrates the importance of reducing not only the energetic losses but also the exergetic losses of the system to a minimum.

References

1. Massmeyer, K. and Posorski, R. (1982) Wärmeübergänge am Energieabsorber und deren Abhängigkeit von meteorologischen Parametern, Institut für Kernphysik, Kernforschungsanlage Jülich GmbH, Jülich.
2. Keller, J. (1985) Characterization of the thermal performance of uncovered solar collectors by parameters including the dependence on wind velocity. Proceedings of the Second Workshop on Solar Assisted Heat Pumps with Ground Coupled Storage, May 1985, Vienna, Austria (ed. D. van Hattem), pp. 367–400.
3. Pitz-Paal, R. (1988) Kondensation an unabgedeckten Sonnenkollektoren. Diploma thesis, Ludwig-Maximilians-Universität München.
4. Soltau, H. (1992) Testing the thermal performance of uncovered solar collectors. *Solar Energy*, **49** (4), 263–272.
5. Morrison, G.L. (1994) Simulation of packaged solar heat-pump water heaters. *Solar Energy*, **53** (3), 249–257.
6. Eisenmann, W., Müller, O., Pujiula, F., and Zienterra, G. (2006) Metal roofs as unglazed solar collectors, coupled with heat pump and ground storage: gains from

condensation, basics for system concepts. Proceedings of the EuroSun 2006 Conference, Glasgow, Scotland, Paper 256.

7. Frank, E. (2007) Modellierung und Auslegungsoptimierung unabgedeckter Solarkollektoren für die Vorerwärmung offener Fernwärmenetze. Ph.D. thesis, Universität Kassel, Fachbereich Maschinenbau.

8. Bertram, E., Glembin, J., Scheuren, J., and Rockendorf, G. (2010) Condensation heat gains on unglazed solar collectors in heat pump systems. Proceedings of the EuroSun 2010 Conference, Graz, Austria.

9. Perers, B. (2010) An improved dynamic solar collector model including condensation and asymmetric incidence angle modifiers. Proceedings of the EuroSun 2010 Conference, Graz, Austria.

10. Perers, B., Kovacs, P., Pettersson, U., Björkman, J., Martinsson, C., and Eriksson, J. (2011) Validation of a dynamic model for unglazed collectors including condensation. Application for standardised testing and simulation in TRNSYS and IDA. Proceedings of the ISES Solar World Congress 2011, August 28–September 2, Kassel, Germany.

11. Bunea, M., Eicher, S., Hildbrand, C., Bony, J., Perers, B., and Citherlet, S. (2012) Performance of solar collectors under low temperature conditions: measurements and simulations results. Proceedings of the EuroSun 2012 Conference, Rijeka and Opatija, Croatia.

12. Palyvos, J. (2008) A survey of wind convection coefficient correlations for building envelope energy systems' modeling. *Applied Thermal Engineering*, **28** (8–9), 801–808.

13. Philippen, D., Haller, M.Y., and Frank, E. (2011) Einfluss der Neigung auf den äusseren konvektiven Wärmeübergang unabgedeckter Absorber. 21st Symposium "Thermische Solarenergie", May 11–13, OTTI Regensburg, Bad Staffelstein, Germany, CD.

14. Perers, B. and Bales, C. (2002) A Solar Collector Model for TRNSYS Simulation and System Testing. A Technical Report of Subtask B of the IEA-SHC – Task 26.

15. Frank, E. and Vajen, K. (2006) Comparison and assessment of numerical models for uncovered collectors. Proceedings of the EuroSun 2006 Conference, Glasgow, Scotland, Paper 256.

16. Isakson, P. and Eriksson, L.O. (1994) MFC 1.0Beta Matched Flow Collector Model for Simulation and Testing – User's Manual, Royal Institute of Technology, Stockholm, Sweden.

17. Haller, M. (2012) TRNSYS Type 832 v5.00 "Dynamic Collector Model by Bengt Perers" – Updated Input–Output Reference, Institut für Solartechnik SPF, Hochschule für Technik HSR, Rapperswil, Switzerland.

18. Carbonell, D., Cadafalch, J., and Consul, R. (2013) Dynamic modelling of flat plate solar collectors. Analysis and validation under thermosyphon conditions. *Solar Energy*, **89**, 100–112.

19. Vela Solaris (2012) Polysun Simulation Software Benutzerdokumentation, Winterthur, Switzerland.

20. Duffie, J.A. and Beckman, W.A. (2006) *Solar Engineering of Thermal Processes*, John Wiley & Sons, Inc., Hoboken, NJ.

21. Ambrosetti, P. and Keller, J. (1985) Das neue Bruttowaermeertragsmodell fuer verglaste Sonnenkollektoren – Teil 1 Grundlagen – Teil 2 Tabellen – 2. überarbeitete Auflage, Eidgenössisches Institut für Reaktorforschung (EIR), Würenlingen.

22. Wemhöner, C. (2011) Field monitoring. Results of field tests of heat pump systems in low energy houses. IEA HPP Annex 32. Economical heating and cooling systems for low energy houses, International Energy Agency Heat Pump Programme.

23. Miara, M., Günther, D., Kramer, T., Oltersdorf, T., and Wapler, J. (2011) Wärmepumpen Effizienz – Messtechnische Untersuchung von Wärmepumpenanlagen zur Analyse und Bewertung der Effizienz im realen Betrieb, Fraunhofer ISE, Freiburg, Germany.

24. Wyssen, I., Gasser, L., and Wellig, B. (2013) Effiziente Niederhub-Wärmepumpen und -Klimakälteanlagen. News aus der Wärmepumpen-Forschung – 19. Tagung des BFE-Forschungsprogramms "Wärmepumpen und Kälte", Burgdorf, Switzerland, pp. 22–35.

25. Oltersdorf, T., Oliva, A., and Henning, H.-M. (2011) Experimentelle Untersuchung eines Direktverdampfers für solar unterstützte Luft/Wasser-Wärmepumpen kleiner Leistung mit zwei Wärmequellen. 21st OTTI Symposium "Thermische Solarenergie", May 11–13, Bad Staffelstein, Germany.

26. Dott, R., Afjei, T. *et al.* (2013) Models of Sub-components and Validation for the IEA SHC Task 44/HPP Annex 38 – Report C2 Part C: Heat Pump Models. A Technical Report of Subtask C – Final Draft.

27. Jin, H. and Spitler, J.D. (2002) A parameter estimation based model of water-to-water heat pumps for use in energy calculation programs. *ASHRAE Transactions*, **108** (1), 3–17.

28. Afjei, T. (1989) YUM: A Yearly Utilization Model for Calculating the Seasonal Performance Factor of Electric Driven Heat Pump Heating Systems, Eidgenössische Technische Hochschule, Zürich.

29. Marti, J., Witzig, A., Huber, A., and Ochs, M. (2009) Simulation von Wärmepumpen-Systemen in Polysun, im Auftrag des Bundesamt für Energie BFE, Bern.

30. Bühring, A. (2001) Theoretische und experimentelle Untersuchungen zum Einsatz von Lüftungs-Kompaktgeräten mit integrierter Kompressionswärmepumpe. Ph.D. thesis, Technical University Hamburg-Harburg.

31. Bertsch, S.S. and Groll, E.A. (2008) Two-stage air-source heat pump for residential heating and cooling applications in northern US climates. *International Journal of Refrigeration*, **31** (7), 1282–1292.

32. Sahinagic, R., Gasser, L., Wellig, B., and Hilfiker, K. (2008) LOREF: Luftkühler-Optimierung mit Reduktion von Eis- und Frostbildung – Optimierung des Lamellenluftkühlers/Verdampfers von Luft/Wasser-Wärmepumpen – Teil 2: Mathematisch-physikalische Simulation des Lamellenluftkühlers mit Kondensat- und Frostbildung, Hochschule Luzern – Technik & Architektur, Horw.

33. Madani, H., Claesson, J., and Lundqvist, P. (2011) Capacity control in ground source heat pump systems. Part I. Modeling and simulation. *International Journal of Refrigeration*, **34** (6), 1338–1347.

34. Heinz, A. and Haller, M. (2012) Appendix A3 – Description of TRNSYS Type 877 by IWT and SPF. Models of Sub-Components and Validation for the IEA SHC Task 44/HPP Annex 38 – Part C: Heat Pump Models – Draft. A Technical Report of Subtask C Deliverable C2.1 Part C.

35. Pärisch, P., Mercker, O., Warmuth, J., Tepe, R., Bertram, E., and Rockendorf, G. (2014) Investigations and model validation of a ground-coupled heat pump for the combination with solar collectors. *Applied Thermal Engineering*, **62** (2), 375–381.

36. Hiller, C.C. and Glicksman, L.R. (1976) Improving Heat Pump Performance via Compressor Capacity Control: Analysis and Test, MIT Energy Lab.

37. Krakow, K.I., Lin, S., and Matsuiki, K. (1987) A study of the primary effects of various means of refrigerant flow control and capacity control on the seasonal performance of a heat pump. *ASHRAE Transactions*, **93** (2), 511–524.

38. Afjei, T. (1993) Scrollverdichter mit Drehzahlvariation. Ph.D. thesis, Eidgenössische Technische Hochschule (ETH), Zürich.

39. Lee, C.K. (2010) Dynamic performance of ground-source heat pumps fitted with frequency inverters for part-load control. *Applied Energy*, **87** (11), 3507–3513.

40. Vargas, J.V. and Parise, J.A. (1995) Simulation in transient regime of a heat pump with closed-loop and on-off control. *International Journal of Refrigeration*, **18** (4), 235–243.

41. Carbonell, D., Cadafalch, J., Pärisch, P., and Consul, R. (2012) Numerical analysis of heat pump models. Comparative study between equation-fit and refrigerant cycle based models. Proceedings of the EuroSun 2012 Conference, Rijeka and Opatija, Croatia.

42. Hornberger, M., E-PUMP: Simulation Tool for Electrical Heat Pumps.

43. Afjei, T. and Wetter, M. (1997) TRNSYS Type. Compressor heat pump including frost and cycle losses. Version 1.1. Model description and implementing into TRNSYS.

44. Afjei, T., Wetter, M., and Glass, A. (1997) Dual-stage compressor heat pump including frost and cycle losses. Version 2.0. Model description and implementation in TRNSYS.

45. Anonymous, TESS Component Libraries – General Descriptions. Available at http://www.trnsys.com/tess-libraries/TESSLibs17_General_Descriptions.pdf (accessed March 18, 2013).

46. Schwamberger, K. (1991) *Modellbildung und Regelung von Gebäudeheizungsanlagen mit Wärmepumpen*, Als Ms. gedr. VDI-Verlag, Düsseldorf.

47. Rawlings, R.H.D. and Sykulski, J.R. (1999) Ground source heat pumps: a technology review. *Building Services Engineering Research and Technology*, **20** (3), 119–129.

48. Huber, A. (2005) Erdwärmesonden für Direktheizung – Phase 1: Modellbildung und Simulation, Zürich.

49. Goffin, P., Ritter, V., John, V., Baetschmann, M., and Leibundgut, H. (2011) Analyzing the potential of low exergy building refurbishment by simulation. Proceedings of Building Simulation 2011: 12th Conference of International Building Performance Simulation Association, November 14–16, Sydney.

50. Ochs, F., Haller, M., and Carbonell, D. (2012) Models of Sub-components and Validation for the IEA SHC Task 44/HPP Annex 38 – Part D: Ground Heat Exchangers. A Technical Report of Subtask C.

51. Hellström, G. (1989) Duct Ground Heat Storage Model – Manual for Computer Code, Department of Mathematical Physics, University of Lund, Sweden.

52. Pahud, D. (1993) Etude du Centre Industriel et Artisanal Marcinhès à Meyrin (GE).

53. Wetter, M. and Huber, A. (1997) TRNSYS Type 451 – Vertical Borehole Heat Exchanger EWS Model Version 2.4, Zentralschweizerisches Technikum Luzern & Huber Energietechnik.

54. Huber, A. and Schuler, O. (1997) Berechnungsmodul für Erdwärmesonden, im Auftrag des Bundesamtes für Energiewirtschaft, Bern.

55. Pahud, D., Fromentin, A., and Hadorn, J.C. (1996) The Duct Ground Heat Storage Model (DST) for TRNSYS Used for the Simulation of Heat Exchanger Piles, DGC-LASEN, Lausanne.

56. Pahud, D. (2007) Serso, stockage saisonnier solaire pour le dégivrage d'un pont.

57. Pahud, D. (2012) The Superposition Borehole Model for TRNSYS 16 or 17 (TRNSBM) – User Manual for the April 2012 Version – Internal Report, Scuola universitaria professionale della Svizzera italiana (SUPSI), Lugano.

58. Eskilson, P. (1986) Superposition Borehole Model – Manual for Computer Code, Department of Mathematical Physics, University of Lund, Sweden.

59. Pahud, D. and Lachal, B. (2005) Mesure des performances thermiques d'une pompe à chaleur couplée sur des sondes géothermiques à Lugano (TI), Bundesamt für Energie.

60. Pahud, D. (2007) PILESIM2: Simulation Tool for Heating/Cooling Systems with Energy Piles or Multiple Borehole Heat Exchangers. User Manual, ISAAC–DACD–SUPSI, Switzerland.

61. Pahud, D. and Hubbuch, M. (2007) Mesures et optimisation de l'installation avec pieux énergétiques du Dock Midfield de l'aéroport de Zürich. Rapport final, Office fédéral de l'énergie, Berne, Suisse.

62. Hellström, G. and Sanner, B. (2000) Earth Energy Designer, User's Manual, Version 2.0.

63. Spitler, J.D. (2000) GLHEPRO – a design tool for commercial building ground loop heat exchangers. Proceedings of the Fourth International Heat Pumps in Cold Climates Conference, Aylmer, Québec, pp. 17–18.

64. Huber, A. and Pahud, D. (1999) Erweiterung des Programms EWS für Erdwärmesondenfelder, im Auftrag des Bundesamtes für Energie, Bern.

65. Ochs, F. and Feist, W. (2012) Experimental results and simulation of ground heat exchangers for a solar and heat pump system for a passive house. Innostock 2012 – 12th International Conference on Energy Storage, Lleida, Spain, INNO-U-23, pp. 1–10.

66. Bianchi, M.A. (2006) Adaptive modellbasierte prädiktive Regelung einer Kleinwärmepumenanlage. Ph.D. thesis, Eidgenössische Technische Hochschule (ETH), Zürich.

67. Lamarche, L., Kajl, S., and Beauchamp, B. (2010) A review of methods to evaluate borehole thermal resistances in geothermal heat-pump systems. *Geothermics*, **39** (2), 187–200.

68. Eskilson, P. (1987) Thermal analysis of heat extraction boreholes. Ph.D. thesis, Department of Mathematical Physics, University of Lund.

69. Yavuzturk, C. and Spitler, J.D. (1999) A short time step response factor model for vertical ground loop heat exchangers. *ASHRAE Transactions*, **105** (2), 475–485.

70. Claesson, J. and Eskilson, P. (1987) Conductive heat extraction by a deep borehole: analytical studies, in *Thermal Analysis of Heat Extraction Boreholes*, Department of Mathematics and Physics, University of Lund, Sweden, pp. 1–26.

71. Lamarche, L. and Beauchamp, B. (2007) A new contribution to the finite line-source model for geothermal boreholes. *Energy and Buildings*, **39** (2), 188–198.

72. Javed, S. and Claesson, J. (2011) New analytical and numerical solutions for the short-term analysis of vertical ground heat exchangers. *ASHRAE Transactions*, **117** (1), 3.

73. Mei, V.C. (1986) Horizontal Ground-Coil Heat Exchanger: Theoretical and Experimental Analysis, Oak Ridge National Laboratory, Oak Ridge, TN, USA.

74. Giardina, J.J. (1995) Evaluation of ground coupled heat pumps for the State of Wisconsin. Master thesis, University of Wisconsin.

75. Glück, B. (2009) Simulationsmodell "Erdwärmekollektor" zur wärmetechnischen Beurteilung von Wärmequellen, Wärmesenken und Wärme-/Kältespeichern, Rud. Otto Meyer-Umwelt-Stiftung, Hamburg.

76. Ramming, K. (2007) Bewertung und Optimierung oberflächennaher Erdwärmekollektoren für verschiedene Lastfälle. Ph.D. thesis, TU Dresden, Dresden, Germany.

77. Hollmuller, P. and Lachal, B. (1998) TRNSYS compatible moist air hypocaust model. Final report, Centre universitaire d'études des problèmes de l'énergie, Genève.

78. Ochs, F., Peper, S., Schnieders, J., Pfluger, R., Janetti, M.B., and Feist, W. (2011) Monitoring and simulation of a passive house with innovative solar heat pump system. ISES Solar World Congress, Kassel.

79. Ochs, F. and Feist, W. (2012) FE Erdreich-Wärmeübertrager Model Für Dynamische Gebäude-und Anlagensimulation mit Matlab/Simulink. BAUSIM 2012, Berlin, Germany.

80. Javed, S. (2012) Thermal modelling and evaluation of borehole heat transfer. Ph.D. thesis, Building Services Engineering, Department of Energy and Environment, Chalmers University of Technology, Göteborg, Sweden.

81. Kjellsson, E., Hellström, G., and Perers, B. (2010) Optimization of systems with the combination of ground-source heat pump and solar collectors in dwellings. *Energy*, **35** (6), 2667–2673.

82. Bertram, E., Pärisch, P., and Tepe, R. (2012) Impact of solar heat pump system concepts on seasonal performance – simulation studies. Proceedings of the EuroSun 2012 Conference, Rijeka and Opatija, Croatia.

83. Bales, C., Gantenbein, P., Jaenig, D., Kerskes, H., and van Essen, M. (2008) Final Report of Subtask B "Chemical and Sorption Storage". The Overview. A Report of IEA SHC Task 32 "Advanced Storage Concepts for Solar and Low Energy Buildings".

84. N'Tsoukpoe, K.E., Liu, H., Le Pierrès, N., and Luo, L. (2009) A review on long-term sorption solar energy storage. *Renewable and Sustainable Energy Reviews*, **13** (9), 2385–2396.

85. Kerskes, H., Mette, B., Bertsch, F., Asenbeck, S., and Drück, H. (2011) Development of a thermo-chemical energy storage for solar thermal applications. Proceedings of the ISES, Solar World Congress Proceedings.

86. Hadorn, J.C. (2005) Thermal Energy Storage for Solar and Low Energy Buildings. State of the Art by the IEA Solar Heating and Cooling Task 32, Servei de publicacions, Universitat de Lleida.

87. Abedin, A.H. and Rosen, M.A. (2011) A critical review of thermochemical energy storage systems. *The Open Renewable Energy Journal*, **4**, 42–46.

88. Duffie, J.A. and Beckman, W.A. (1996) *Solar Engineering of Thermal Processes*, 3rd edn, John Wiley & Sons, Inc., Hoboken, NJ.

89. Shah, S.A., Short, T.H., and Peter Fynn, R. (1981) A solar pond-assisted heat pump for greenhouses. *Solar Energy*, **26** (6), 491–496.

90. Hughes, P.J., Klein, S.A., and Close, D.J. (1976) Packed bed thermal storage models for solar air heating and cooling systems. *Journal of Heat Transfer (United States)*, **98** (2), 336–338.

91. Novo, A.V., Bayon, J.R., Castro-Fresno, D., and Rodriguez-Hernandez, J. (2010) Review of seasonal heat storage in large basins: water tanks and gravel–water pits. *Applied Energy*, **87** (2), 390–397.

92. Bauer, D., Marx, R., Nußbicker-Lux, J., Ochs, F., Heidemann, W., and Müller-Steinhagen, H. (2010) German central solar heating plants with seasonal heat storage. *Solar Energy*, **84** (4), 612–623.

93. Haller, M.Y. (2010) Combined solar and pellet heating systems – improvement of energy efficiency by advanced heat storage techniques, hydraulics, and control. Ph.D. thesis, Graz University of Technology, Graz, Austria.

94. Van Berkel, J. (1997) Thermocline entrainment in stratified energy stores. Ph.D. thesis, Technical University Eindhoven.

95. Shah, L.J. and Furbo, S. (1998) Correlation of experimental and theoretical heat transfer in mantle tanks used in low flow SDHW systems. *Solar Energy*, **64** (4–6), 245–256.

96. Drück, H. (2007) Mathematische Modellierung und experimentelle Prüfung von Warmwasserspeichern für Solaranlagen. Ph.D. thesis, Institut für Thermodynamik und Wärmetechnik (ITW) der Universität Stuttgart, Shaker Verlag, Aachen.

97. Nizami, D. (2010) Computational fluid dynamics study and modelling of inlet jet mixing in solar domestic hot water tank systems. MASc thesis, McMaster University, Ontario, Canada.

98. Logie, W. and Frank, E. (2011) A computational fluid dynamics study on the accuracy of heat transfer from a horizontal cylinder into quiescent water. Proceedings of the ISES Solar World Congress 2011, August 28–September 2, Kassel, Germany.

99. Allard, Y., Kummert, M., Bernier, M., and Moreau, A. (2011) Intermodel comparison and experimental validation of electrical water heater models in TRNSYS. Proceedings of Building Simulation 2011 – 12th Conference of the IBPSA, Sydney, Australia.

100. Zurigat, Y.H., Maloney, K.J., and Ghajar, A.J. (1989) A comparison study of one-dimensional models for stratified thermal storage tanks. *Journal of Solar Energy Engineering*, **111** (3), 204–210.

101. Zurigat, Y.H., Liche, P.R., and Ghajar, A.J. (1991) Influence of inlet geometry on mixing in thermocline thermal energy storage. *International Journal of Heat and Mass Transfer*, **34** (1), 115–125.

102. Leibfried, U., Günzl, A., and Sitzmann, B. (2008) SOLAERA: Solar-Wärmepumpe-system im Feldtest. 18th OTTI Symposium "Thermische Solarenergie", Bad Staffelstein, Germany.

103. Loose, A., Bonk, S., and Drück, H. (2012) Investigation of combined solar thermal and heat pump systems – field and laboratory tests. Proceedings of the EuroSun 2012 Conference, Rijeka and Opatija, Croatia.

104. Philippen, D., Haller, M.Y., Logie, W., Thalmann, M., Brunold, S., and Frank, E. (2012) Development of a heat exchanger that can be de-iced for the use in ice stores in solar thermal heat pump systems. Proceedings of the EuroSun 2012 Conference, Rijeka and Opatija, Croatia.

105. Mehling, H. and Cabeza, L.F. (2008) *Heat and Cold Storage with PCM*, Springer.

106. Sharma, A., Tyagi, V.V., Chen, C.R., and Buddhi, D. (2009) Review on thermal energy storage with phase change materials and applications. *Renewable and Sustainable Energy Reviews*, **13** (2), 318–345.

107. Regin, A.F., Solanki, S.C., and Saini, J.S. (2009) An analysis of a packed bed latent heat thermal energy storage system using PCM capsules: numerical investigation. *Renewable Energy*, **34** (7), 1765–1773.

108. Heinz, A. and Streicher, W. (2006) Application of phase change materials and PCM slurries for thermal energy storage. Proceedings of the Ecostock 2006, Stockton.

109. Brousseau, P. and Lacroix, M. (1996) Study of the thermal performance of a multi-layer PCM storage unit. *Energy Conversion and Management*, **37** (5), 599–609.

110. Simard, A.P. and Lacroix, M. (2003) Study of the thermal behavior of a latent heat cold storage unit operating under frosting conditions. *Energy Conversion and Management*, **44** (10), 1605–1624.

111. Stritih, U. and Butala, V. (2003) Optimisation of thermal storage combined with biomass boiler for heating buildings. 9th International Conference on Thermal Energy Storage, Warsaw.

112. Ibáñez, M., Cabeza, L.F., Solé, C., Roca, J., and Nogués, M. (2006) Modelization of a water tank including a PCM module. *Applied Thermal Engineering*, **26** (11–12), 1328–1333.

113. Bony, J. and Citherlet, S. (2007) Numerical model and experimental validation of heat storage with phase change materials. *Energy and Buildings*, **39** (10), 1065–1072.

114. Dincer, I. and Rosen, M.A. (2002) *Thermal Energy Storage: Systems and Applications*, John Wiley & Sons, Ltd, West Sussex, UK.

115. Schranzhofer, H., Puschnig, P., Heinz, A., and Streicher, W. (2006) Validation of a TRNSYS simulation model for PCM energy storages and PCM wall construction elements. Proceedings of the Ecostock 2006, Stockton.

116. Puschnig, P., Heinz, A., and Streicher, W. (2005) TRNSYS simulation model for an energy storage for PCM slurries and/or PCM modules. Second Conference on Phase Change Material & Slurry: Scientific Conference & Business Forum, Yverdons-les-Bains, Switzerland.

117. Streicher, W. (2008) Simulation Models of PCM Storage Units. A Report of IEA Solar Heating and Cooling Programme. Task 32 "Advanced Storage Concepts for Solar and Low Energy Buildings". Report C5 of Subtask C, Graz, Austria.

118. Behschnitt, S. (1996) TRNSYS Type 207 – Ice Storage Tank. Available at http://sel.me.wisc.edu/trnsys/trnlib/library16.htm (accessed April 9, 2014).

119. SEL, CSTB, TRANSSOLAR, and TESS (2012) TRNSYS 17 – A TRansient SYstem Simulation Program.

120. Drück, H. (2006) Multiport Store Model for TRNSYS – Type 340 – V1.99F.

121. Haller, M.Y. and Frank, E. (2011) On the potential of using heat from solar thermal collectors for heat pump evaporators. Proceedings of the ISES Solar World Congress 2011, August 28–September 2, Kassel, Germany.

122. Pärisch, P., Warmuth, J., Bertram, E., and Tepe, R. (2012) Experiments for combined solar and heat pump systems. Proceedings of the EuroSun 2012 Conference, Rijeka and Opatija, Croatia.

123. Hollands, K.G.T. and Lightstone, M.F. (1989) A review of low-flow, stratified-tank solar water heating systems. *Solar Energy*, **43** (2), 97–105.

124. Huggenberger, A. (2013) Schichtung in thermischen Speichern – Konstruktive Massnahmen am Einlass zum Erhalt der Schichtung. Bachelor thesis, Hochschule für Technik HSR, Institut für Solartechnik SPF, Rapperswil, Switzerland.

4 Performance and its assessment

Ivan Malenković, Peter Pärisch, Sara Eicher, Jacques Bony, and Michael Hartl

> **Summary**
>
> For a transparent performance assessment of any energy conversion system, clear definitions of system boundaries for energy balancing and performance figures are essential. Generally, depending on the aim of the system evaluation, different figures and energy balances will be chosen. This chapter proposes a set of system boundaries and performance figures for the energetic and environmental performance evaluation of solar and heat pump (SHP) systems. Given definitions, however, can principally be applied on any system providing heating, cooling, and domestic hot water (DHW), thus providing a basis for transparent comparisons among different technologies.
>
> In Section 4.2, different performance figures for heat pumps and solar thermal collectors and systems are presented and discussed. A set of system boundaries for the analysis and performance evaluation of SHP systems, as well as corresponding performance figures, is given in Section 4.3. Environmental evaluation of SHP systems is discussed in Section 4.4. Finally, in Section 4.5 the usage of all performance figures defined in previous sections is explained in form of a calculation example.

4.1 Introduction

Performance of a solar and heat pump system or any energy conversion system for heating or cooling applications in general is often being simplified and equalized with its efficiency, whereas this efficiency is frequently expressed as a single number without any proper description of the operating conditions or system boundaries used for its calculation. The performance of a system, however, can also be regarded as a broader concept including energy-, economy-, and environment-related aspects of system operation within certain boundaries under given operating conditions for a defined time period.

Knowledge about the system performance is important in many ways: by means of an energy performance analysis, an optimization of the system can be carried out and a product or a part of it improved, for example; an analysis of the economy of the product can provide important information for the customer and help him/her to decide on the appropriate product or technology for the specific application; and an environmental analysis can provide information on the depletion of nonrenewable energy sources and greenhouse gas emissions, which may help establishing adequate energy policy, suitable and target-oriented research funding, or incentives for manufacturers and/or end users on national and international levels.

Solar and Heat Pump Systems for Residential Buildings, First Edition.
Edited by Jean-Christophe Hadorn.
© 2015 Ernst & Sohn GmbH & Co. KG. Published 2015 by Ernst & Sohn GmbH & Co. KG.

With such a variety of goals, it can be very difficult or even impossible to provide only one performance figure, which gives enough information to all target groups and is applicable to all technologies. On the other hand, it would be favorable to have a minimum of clearly defined, broadly applicable, and consistent set of system boundaries, performance figures, and reporting procedures, which would allow for a fair, transparent, and comparable reporting on different systems and technologies, their analysis, and comparison regarding previously mentioned aspects. Within T44A38, a methodology for defining intercomparable performance figures and system boundaries was elaborated in cooperation with similar activities in the field of other technologies involving solar thermal and heat pumps (IEE Project QAiST – Quality Assurance in Solar Heating and Cooling Technologies – www.qaist.org; IEA HPP Annex 34 – Thermally Driven Heat Pumps for Heating and Cooling – www.annex34.org).

System boundaries used for performance evaluation are arbitrary and can be chosen according to the assessment method, the aim of performance evaluation, available data, and so on. The boundaries can stretch both on the spatial scale and on the timescale. This means that different system boundaries not only can include a different number of system components and interfaces with its environment, but can also consider different timescales within the lifetime of the system for its evaluation: it can take into consideration only the system interaction with its environment during the operation or can include the energy and material consumption, as well as their consequences, needed for the production, transport, and installation of the system, for example.

Depending on the goal of the performance evaluation, different system boundaries will be used and the choice of performance figures for reporting may vary. Whatever the aim and the choice, it is crucial to clearly define both the system boundaries and the performance figures used in order to avoid misunderstandings and to provide intelligible information for the user. This may be done either by an adequate definition of both within the publication, report, or technical manual or by referencing the used figure to a standard or other relevant document (e.g., guideline or publication). In addition, in order to be able to assess the quality of the system and to put the reported figures into a perspective, it is often necessary to provide information on the operating conditions such as climate, supply temperature, and average solar irradiation. For standardized evaluation methods, such as laboratory tests, these data are often included in the method itself and a reference to a document might suffice. Otherwise, it can be very difficult to choose the right amount of information to be provided to give a full picture on one hand and to avoid overloading on the other hand.

Performance assessment of a system can be in general either a forecast based on certain assumptions regarding the system itself and its environment or an evaluation of measurement data from the system operation – either from the field trials under real operating conditions or from defined and controlled laboratory tests. For the performance prediction of SHP systems, methods currently used involve numerical simulation and laboratory tests, as well as their combinations. Until now, simple calculation methods based on laboratory tests on single system components and simplified climate data (such as the temperature bin method used, for example, in

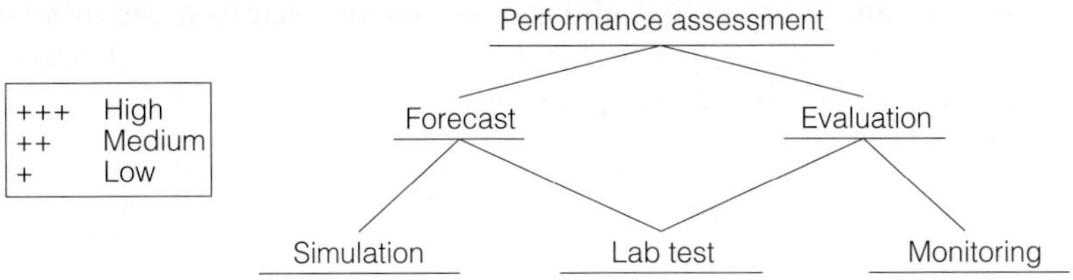

4.2 Definition of performance figures

		Simulation	Lab test	Monitoring
+++ High ++ Medium + Low				
Complexity		+++	+++	++
Infrastructure requirement		++	+++	+/++
Required time		+/++	++	+++
Required skills		+++	++	++
Accuracy		++	++/+++	++/+++
Required knowledge about system		+++	++	++
Required knowledge about components		+++	+	+
Suitability for system analysis		+++	++/+++	+/+++
Comparability		+++	++/+++	+
Repeatability		+++	+++	+
Suitability for labeling		++	+++	+

Performance assessment branches into Forecast (Simulation, Lab test) and Evaluation (Lab test, Monitoring).

Fig. 4.1 Different methods for SHP system assessment with basic characteristics

heat pump standards EN 14825 or EN 15316-4-2) have not been available. For the performance assessment of an installed system, measurements on the system itself and its interfaces to the energy sources and sinks are necessary, whereas the effort for data acquisition and processing may vary substantially depending on the system complexity and extent of analysis. An overview of different assessment methods and their characteristics is given in Figure 4.1.

In the following sections, recommendations for the performance assessment of SHP systems regarding their energy and environmental performance are given. Economic issues are discussed in Chapter 8.

4.2 Definition of performance figures

4.2.1 Overview of performance figures in current normative documents

A variety of national and international standards and other normative documents are available for both heat pump and solar thermal technologies. These documents cover component and system testing, as well as calculation methods for their performance prediction. In most of them, performance figures are defined, both for components and for systems containing more than one component (e.g., heat pump and storage).

Analyzing these documents, however, it soon becomes obvious that there are different definitions of the same performance figure or that the same energy balance defining a performance figure is indicated in a different manner.

Within T44A38, a number of significant normative documents for both heat pumps and solar thermal collectors were collected and analyzed. From this analysis, a coherent nomenclature system for different performance figures was defined, which was applied on the systems investigated within the activity. This nomenclature is used throughout the Task/Annex, if not stated differently. For example, in some cases, system performance was evaluated before the definition of the nomenclature. In these cases, system boundaries might differ from the ones described here.

As at the time of writing the book, no normative documents for testing of combined solar thermal and heat pump systems were available, mainly documents covering single technologies were analyzed. However, within the European Commission's Directives on ecodesign and ecolabeling of energy using products, calculation methods for the energy efficiency of space heating devices, including combinations thereof, have been published. A current status is given in Section 4.2.2.

4.2.1.1 Heat pumps

National and international standards and guidelines for heat pump testing and performance assessment are well developed and widely applied in all important markets worldwide. In many regions, they are commonly used as a basis for marketing tools such as quality labels and subsidies. They are available for heating, domestic hot water, and cooling applications. Generally, the performance for each application is being calculated separately. Some documents, however, contain figures that take into account both heating and domestic hot water production. The performances for simultaneous heating and cooling or cooling and DHW are not covered in any of the collected documents.

The list of collected and reviewed standards and guidelines, as well as relevant nomenclature and definitions of performance figures, is provided in Appendix 4.A.1. From these documents, some general conclusions regarding the nomenclature and definition of specified performance figures can be made:

Coefficient of performance (COP): In all reviewed standards, it is used for the performance of the heat pump unit under stationary operating conditions. An exception is the EN 16147, where the system heat pump with DHW storage is considered and the figure is calculated for a set of different, partly transient operating conditions. The influence of the liquid circulation pumps on the energy inputs and outputs differs among the standards. It is generally used for heating applications only.

Energy efficiency ratio (EER): Same as COP, but used for cooling applications in most of the European standards. In many US standards, it is defined as Btu/h of cooling energy per watt of electricity consumption.

Seasonal coefficient of performance (SCOP): Used only in one standard to express the calculated seasonal heating efficiency of the heat pump for assumed boundary conditions, for example, climate and building load.

4.2 Definition of performance figures

Seasonal energy efficiency ratio (SEER): Same as SCOP, but for cooling applications.

Seasonal performance factor (SPF): In European standards (e.g., EN 15316-4-2), the SPF is used as a figure to express the efficiency of the overall system including all auxiliary components such as circulation pumps, storages, and backup heaters. In the VDI 4650 guideline, it is used to express roughly the same efficiency as the SCOP – not taking into account the whole system but only the heat pump unit with some auxiliary energy. Its equivalent in the US standards is the heating SPF (HSPF).

4.2.1.2 Solar thermal collectors

Unlike the heat pumps, testing and performance evaluation standards are available both for collectors and for whole systems. References to the documents and a short overview are given in Appendix 4.A.2.

There is a high level of agreement between the reviewed standards for solar thermal applications regarding the definition of performance figures and used nomenclature. Four main performance indicators are found:

Collector thermal efficiency (η): A component performance figure, it represents the ratio of the heat energy provided by the collector to the product of a defined collector area and the incident solar irradiation on the collector for a certain time period under steady- and non-steady-state conditions.

Solar fraction (f_{sol}): A system performance figure representing the ratio of the thermal energy provided by the solar system to the total system load for the same period of time, mostly a season or a year.

Fractional energy savings (f_{sav}): Also a system performance figure expressing the potential reduction of purchased energy by the usage of solar system, compared with a reference system under the assumption that both systems give the same thermal comfort over the same specified period.

Thermal performance: In the reviewed standards, no system performance figure similar to SPF for the heat pump systems was found. However, from the delivered heat and the parasitic energy, as defined in the reviewed standards, the efficiency of the system could be calculated in the same manner as for the heat pump systems.

4.2.2 Solar and heat pump systems

At the time this book was written, the European Union already established the "European Union Ecodesign Directive for Energy-Related Products" [1]. In it, different types of energy-related products are arranged in lots according to their purpose. Devices for space heating and combined space heating and domestic hot water preparation with heating capacity lower than 400 kW are covered by lot 1 and pure domestic hot water heaters and hot water storage tanks are covered by lot 2.

The Directive provides rules for labeling of products for heating and domestic hot water preparation similarly to the label already known from various electric home appliances. The products with classes between A+++ and G – with A+++ for the most efficient and G for the least efficient products – make energy consumption of the heating installations transparent and motivate the consumer to focus on energy-efficient products. Appendix 4.A.3 contains an overview of the most important documents related to that directive.

The classification is made according to the calculated energy efficiency. The energy efficiency (η_s) is calculated according to the Ecodesign regulation and the related standards. η_s is closely related to the primary energy ratio (PER) described in Section 4.4.1.1. Equation 4.1 gives the energy efficiency for heat pumps with SCOP determined according to EN 14825 following the European Commission Regulation No. 813/2013.

$$\eta_s = \frac{\text{SCOP}}{\text{CC}} - F_i = \text{PER}. \tag{4.1}$$

In Equation 4.1, the conversion coefficient (CC) is 2.5 for electricity used for mechanical (vapor) compression heat pumps. The correction factors F_i are expressed in percentage and account for negative contribution to the seasonal space heating energy efficiency of heaters mainly due to the system controls. Their calculation method is still in a draft version.

The need for the labeling of combined systems, as investigated in this book, has been recognized by the European Commission. Therefore, the label for the "packages of combination heater, temperature control, and solar devices" was introduced to the EU Regulation No. 811/2013. This label allows for accounting additional heat input from supplementary space heating devices such as solar devices. However, the method is very simple and there is a need to investigate whether it can be used for fair and consistent evaluation of SHP systems considering the variety and complexity of existing products and applications.

4.2.3 Efficiency and performance figures

Efficiency is one of the main performance indicators for heating and cooling systems and is generally defined as the ratio between the useful energy output from the system to the energy input to the system. The energy input is commonly limited to the driving energy needed to transform the primary (including environmental) and/or final energy into useful energy and deliver it to the end user. As already pointed out previously, depending on the aim of the performance assessment of an energy conversion system, the interesting aspects might be, for instance, economy-, environment-, or energy-related evaluation of the system operation. In this section, the energy-related evaluation of system performance will be discussed; the environmental aspects will be highlighted in Section 4.4.

The efficiency of a heating or a cooling system that is, in general, an energy conversion system using one or more driving energy inputs to produce heating and/or cooling effect or useful energy output is expressed as a ratio of these energy flows to and from the system (Equation 4.2). Note that even though in cooling applications the removed energy from the user is added to the cooling system, this energy is still regarded as a "useful energy output" of the system and positive values are used in the equations.

$$\eta_{\text{sys}} = \frac{\sum_{i=1}^{n} Q_{\text{out},i}}{\sum_{j=1}^{m} Q_{\text{in},j}} = \frac{Q_{\text{out}}}{Q_{\text{in}}}. \tag{4.2}$$

4.2 Definition of performance figures

Q_{out} contains the considered useful energy provided by the system while Q_{in} represents the sum of all considered energy inputs to the system during the chosen time period. As discussed in Section 4.4, Q_{out} and Q_{in} will generally differ depending on the choice of system boundaries yielding generally different definitions of performance figures.

Another distinction that can be made between different performance figures is according to the operating conditions under which they were determined. For heat pumps, as for other energy conversion units, a distinction between steady-state operating conditions, mostly provided under controlled laboratory conditions, and transient conditions immanent to real operating environments in the field is often made. As described in Chapter 5, a trend toward transient measurement under controlled laboratory conditions, especially for complex systems, could be observed at the time of writing this book. In addition to the time profile of the operating conditions, it is essential to give information on other parameters such as heat sink or heat source temperature levels for the heat pump or solar irradiation levels for the collector to provide full information to the user. By just stating a performance figure without the conditions under which it was determined, room for misinterpretation is given.

A difference between the component and system performance figures is often being made. In many cases, however, this difference is arbitrary and depends on the definition of the system boundaries. Thus, for example, as a compressor can be a component of a heat pump unit, the heat pump unit can represent a component of an SHP system. In the following, the heat pump unit, the solar collector, or the backup unit will be regarded as components – parts of an SHP system or one of its subsystems.

It is also necessary to exactly define the "energy quality" involved in the calculation. The energy quality refers to the primary energy content of the energy input or the energy output. Very often, the terms primary energy or useful energy are quite arbitrary. Therefore, clear definitions for these and some other terms are needed. This will be discussed in Section 4.4, where environmental performance evaluation is addressed.

Reporting on performance

When reporting on a performance of a component or a system using usually one performance figure such as COP, η_{coll}, or SPF, the circumstances under which the figure was obtained must also be provided to the reader. Otherwise, a η_{coll} of 98% or an SPF of 2.3 cannot be assessed qualitatively; that is, the reader – customer, planner, or installer – cannot decide whether the system or a component is good or not compared with other products. In many cases, this can lead to misunderstandings and false application of the product. Information on the operation conditions, under which the performance was evaluated, can be given in form of a reference to a normative document or as a set of relevant ambient and operating conditions. The more information is provided about the operating conditions and circumstances during performance evaluation, the easier for the reader to fully understand the meaning of the value of performance figures!

4.2.4 Component performance figures

4.2.4.1 Coefficient of performance

The COP of the heat pump is the ratio between its heating capacity and the overall electricity consumption, both measured under steady-state operating conditions. The counterpart of the COP for cooling applications is the energy efficiency ratio, EER. The system boundary for the energy balancing corresponds to boundary HP in Figure 4.8. Hence, the COP can be calculated as

$$\text{COP} = \frac{\overline{\dot{Q}_{\text{HP,H}}}}{\overline{P_{\text{el,HP}}}}. \tag{4.3}$$

In European standards (e.g., EN 14511-3 or EN 15879-1), the heating (or cooling) capacity for hydronic distribution systems is corrected for the (assumed) amount of energy dissipated from the liquid circulation pump to the heat transfer fluid. The corrective amount of heat is calculated from the measured pressure drop over the heat exchanger and assumed circulation pump efficiency. The electricity consumption due to pumping work needed to overcome the pressure losses of the heat transfer fluid within the heat pump unit is corrected for the same amount of energy.

4.2.4.2 Seasonal coefficient of performance

The SCOP (and its counterpart for the cooling applications – SEER) is efficiency figure calculated from the laboratory measurement data by assuming defined time-dependent operating conditions over a certain period of time. The time dependence considers in general oscillating ambient and heat source temperatures, changing heat supply temperatures, operation under part-load conditions, and so on. The system boundary is the same as for the COP – boundary HP in Figure 4.8.

According to current version of the European standard EN 14825, the SCOP and the SEER are calculated using the temperature bin method. This method is based on the division of the cumulative annual frequency of the outside dry-bulb temperature into temperature classes called bins. For every bin, an average operating condition is defined. It is then assumed that the heat pump unit operates under this condition for the entire temperature range covered by the bin. Finally, the energy inputs, including standby consumption, as well as useful energy outputs over all bins are summed up and the efficiency is calculated. A full description of the method, as well as some open questions, can be found, for example, in Ref. [2].

Within EN 14825, the SCOP and the SEER are defined as follows:

$$\text{SCOP} = \frac{Q_H}{(Q_H/\text{SCOP}_{\text{on}}) + W_{\text{el,off}}}, \tag{4.4}$$

$$\text{SEER} = \frac{Q_C}{(Q_C/\text{SEER}_{\text{on}}) + W_{\text{el,off}}}. \tag{4.5}$$

In the above equations, Q_H and Q_C represent the annual heating and cooling loads, respectively. SCOP_{on} and SEER_{on} are the figures representing efficiencies only for the periods of operation when the heat pump unit and in case of $SCOP_{on}$ the direct electrical backup heating are delivering/extracting useful energy. $W_{el,off}$ is the electricity consumption of the unit in idle state, for example, in standby mode.

4.2.4.3 Solar collector efficiency

The stationary collector efficiency is expressed as the thermal output of the collector $\dot{Q}_{gain}$ divided by the irradiance G_g on the collector pane A_{coll}:

$$\eta_{coll} = \frac{\dot{Q}_{gain}}{G_g \cdot A_{coll}}. \tag{4.6}$$

Standards (e.g., EN 12975-2) define the standard testing conditions, for example, minimal values for irradiance, ambient air temperature, wind speed, and the test procedure (steady state or transient). For covered collectors, the steady-state collector efficiency is described as quadratic equation depending on the temperature difference between collector mean fluid temperature and ambient air temperature. For uncovered collectors, the stationary collector efficiency is described as linear equation regarding the temperature difference and considers wind speed and sky temperature for long-wave irradiance losses.

The efficiency of the solar thermal system is defined as the ratio of the obtained useful heat divided by the irradiation [3] on the collector pane. Depending on how the useful heat is defined and where it is measured, stagnation periods, pipe losses, actual weather conditions, and interdependence on the conventional heating system may be taken into account. Thus, using the nomenclature of this handbook for the useful heat, the collector utilization ratio ω_{SC} may be defined as

$$\omega_{SC} = \frac{Q_{SC,H}}{\int G_g \cdot A_{coll} \, dt}. \tag{4.7}$$

Analogously to heat pump systems, a seasonal performance factor can also be defined (Equation 4.8). Attention has to be paid that within T44A38 $P_{el,SC}$ considers only direct energy consumption of the solar collector like a fan for hybrid collectors. For standard solar collectors, SPF_{SC} is not defined as the energy consumption for the circulation pump $P_{el,SC,H}$ is not considered.

$$SPF_{SC} = \frac{Q_{SC,H}}{\int P_{el,SC} \, dt}. \tag{4.8}$$

4.2.5 System performance figures

4.2.5.1 Seasonal performance factor

In reviewed normative documents and in common practice, the seasonal performance factor is mainly used as a system performance figure, although in some cases it can also be used to express the efficiency of a heat pump unit (e.g., in VDI 4650). The SPF gives the final energy efficiency of the whole system or a defined subsystem, calculated as the overall useful energy output to the overall driving final energy input for an adopted system boundary (Equation 4.9). It expresses the performance of a system over a year or a season – a heating or a cooling season, for example. The same definition can also be used for shorter time periods, such as a week or a month, but a different nomenclature for the figure has to be used, that is, weekly performance figure, for example.

$$\text{SPF} = \frac{\int \left(\dot{Q}_{\text{SH}} + \dot{Q}_{\text{DHW}} + \dot{Q}_{\text{C}}\right) dt}{\int \sum P_{\text{el}} \, dt}. \tag{4.9}$$

In addition to an overall SPF that provides the efficiency of the system in all operation modes, separate SPFs for single operation modes (e.g., heating only, heating and DHW, and cooling and DHW) can be defined. However, if, for example, heating and cooling or heating and DHW are produced simultaneously, it might be difficult to quantify the specific input and/or output energy for a single mode of operation. This has to be taken into account when evaluating the measurement or simulation results. Although the useful cooling energy removed from the system surrounding Q_C has the opposite algebraic sign to the useful heating energy supplied by the system ($Q_{\text{SH}} + Q_{\text{DHW}}$) in the physical sense, for the definition of the performance figures throughout the chapter Q_C is treated as a useful energy output of the system in the engineering manner; that is, positive values are to be used in the equations, as is commonly done in engineering praxis.

The SPF quantifies the efficiency of a system or a subsystem for given operating conditions such as heat source temperature, solar irradiation, and supply temperature profile. Final electrical energy provided on site is considered as energy input. This does not take into consideration the "quality" of the driving energy, for example, in terms of the depletion of nonrenewable resources or greenhouse gas emissions caused during the lifetime of a system. For such an environmental performance of the system, available energy mixes on site have to be taken into account. For that reason, primary energy ratio for nonrenewable part of energy input (PER_{NRE}) and equivalent warming impact of the system (EWI_{sys}) are introduced (Section 4.4).

Note that for non-monoenergetic systems (e.g., if the backup heating system runs on biomass, gas, etc.), the SPF cannot fully reflect the total energy consumption, since it considers only electricity as energy input, as defined for SHP systems. In that case, the consumption of any additional fuel should be given separately and the overall system performance expressed by the primary energy ratio.

Also note that for air source heat pumps, defrosting has to be considered:

- *Direct electric defrosting:* electricity consumption should be included in $P_{\text{el,HP}}$.
- *Hot gas defrosting:* the energy consumption should also be included in $P_{\text{el,HP}}$.
- *Reverse cycle defrosting:* the heat energy taken from the storage/building has to be subtracted from the useful energy output at the appropriate boundary, if not automatically executed by the heat meter.

4.2.6 Other performance figures

4.2.6.1 Solar fraction

The solar fraction specifies the share of energy delivered to the thermal storage by the solar part of the system. Both the solar contribution and the overall output of the system

4.2 Definition of performance figures

need to be precisely defined. For solar and heat pump systems, only the direct solar heat delivered to the conventional part of the system is accounted for as solar heat. Some definitions exist in several standards and publications, which vary in the treatment of the thermal losses of the heat storage.

The first definition from ISO 9488 or EN 12976-2 calculates the ratio of the direct solar heat to the useful heat (Equation 4.10). The storage losses are not considered in that case, which results in the highest values for the solar fraction. In solar active houses with a large storage volume on the heat sink side, very high solar fractions of 1 or even higher could be achieved. However, this does not mean that no additional heat is needed. Higher storage losses increase the demand for direct solar heat as well as for additional heat, which leads to higher solar fractions.

$$f_{sol,1} = \frac{Q_{SC,H}}{Q_{DHW} + Q_{SH}}. \tag{4.10}$$

Contrary to the first definition, in the second one [4] all thermal losses of the storage are subtracted from the direct solar heat, leading to very low values of solar fraction (Equation 4.11). In this case, it is limited to 1, reached when no additional heat is delivered. It can also become negative when the losses exceed the solar contributions for small-scale systems. Higher storage losses that increase the need for additional heat lead to lower solar fractions.

$$f_{sol,2} = \frac{Q_{SC,H} - Q_L}{Q_{DHW} + Q_{SH}} = 1 - \frac{Q_{HP,H} + Q_{BU,H}}{Q_{DHW} + Q_{SH}}. \tag{4.11}$$

The third definition comes from VDI 6002-1 [3]. In this definition, only the solar quota of the storage losses is subtracted from the direct solar heat and then related to the useful heat (Equation 4.12). The definition can be transformed to the ratio of direct solar heat to the total heat production, thus the sum of direct solar heat and all other heat inputs from the system. If there is no additional heat input, the solar fraction becomes 1. Increasing the storage losses leads to a moderate rise of the solar fraction.

$$f_{sol,3} = \frac{Q_{SC,H} - Q_L[Q_{SC,H}/(Q_{HP,H} + Q_{SC,H} + Q_{BU,H})]}{Q_{DHW} + Q_{SH}} \hat{=} \frac{Q_{SC,H}}{Q_{HP,H} + Q_{SC,H} + Q_{BU,H}}. \tag{4.12}$$

The deviations between the three definitions become obvious for a solar active house with a large hot water storage. Figure 4.2 shows the solar fractions for a simulation study according to Task 32 boundary conditions [5]. It is a single-family house in Zurich with a space heating demand of 60 kWh/(m² year) and a DHW consumption of 200 l/day at 45 °C. The collector area of 30 m² is kept constant while the storage volume rises from 1 to 25 m³. Storage losses, conventional heat, and solar heat increase with storage size.

Different definitions for solar fraction yield different values and show principally different trends with increasing storage volume (Figure 4.2). While $f_{sol,2}$ shows an optimal size of the storage volume focusing on the conventional heat consumption, the solar fractions calculated according to the other two definitions ($f_{sol,1}$ and $f_{sol,3}$) increase with storage size. As $f_{sol,1}$ is not limited to 100%, it rises very steeply over the storage volume.

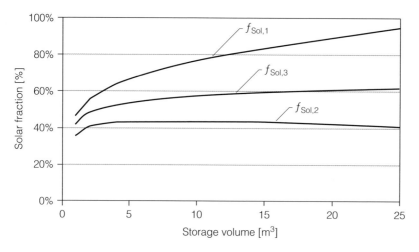

Fig. 4.2 Different definitions of the solar fraction of a solar active house

4.2.6.2 Renewable heat fraction

An SHP system can generally use more than one renewable energy source. In addition to direct usage of solar energy, as previously described, the heat pump may use energy from the ambient air, ground, groundwater, or the solar part of the system as the heat source. In this context, the solar fraction does not reflect the full potential of the system to use renewable energy sources and can thus be seen as a less important figure.

Instead, the renewable heat fraction can be defined as

$$f_{\text{ren,SHP}} = 1 - \frac{\int \left(\sum P_{\text{el,SHP}}\right) dt}{Q_{\text{DHW}} + Q_{\text{SH}}} \hat{=} 1 - \frac{1}{\text{SPF}_{\text{SHP}}}. \tag{4.13}$$

As the renewable heat fraction can be derived from the SPF, it is therefore important to put the system boundary to the index.

4.2.6.3 Fractional energy savings

The fractional energy savings describe the influence of a specific optimization such as the combination with a solar thermal system in relation to a reference system. The choice of the reference system with its properties has a big influence on this performance figure. Penalties, which eventually occur, are added to the electricity consumption of the solar-assisted system. See, for example, Deliverables of Subtask C for their calculation. The system boundary needs to be specified, for example, here SHP. Penalties for the reference system should actually be zero otherwise they need to be added to the electricity consumption of the reference system.

$$f_{\text{sav,SHP}} = 1 - \frac{\left(W_{\text{el,SHP}} + W_{\text{penalty}}\right)_{\text{solar}}}{\left(W_{\text{el,SHP}}\right)_{\text{ref}}} = 1 - \frac{(\text{SPF}_{\text{SHP}})_{\text{ref}}}{(\text{SPF}_{\text{SHP}})_{\text{solar}}}. \tag{4.14}$$

This figure applies only for monoenergetic systems. If there is another energy carrier than electricity, for example, for the backup heater, the primary energy factors have to be taken into account. See Section 4.4.1.3 for fractional primary energy savings.

4.3 Reference system and system boundaries

4.3.1 Reference SHP system

The basis for the energy performance evaluation of an energy transformation system, in the common sense of the term, is energy balancing, that is, summing up its energy inputs and outputs and establishing a relationship between them. As a great variety of different SHP system configurations are available on the market (see Chapter 2), a comparison of the system performances among them might present a challenge. This is even more so, if the systems are not considered just as "black boxes" but a deeper analysis is aimed at.

For every transparent and reliable performance comparison among different systems, it is necessary to define comparable system boundaries for energy balancing – that is, an exact definition of relevant energy inputs and outputs to and from the system, which should be as much independent regarding the system configuration as possible. For that purpose, T44A38 proposed a "reference system" that ideally contains all component arrangements and energy flows of the known configurations. System boundaries for performance evaluation and analysis were defined for that system. The boundaries for particular configurations can then be obtained by removing nonexisting components and energy flows from the reference system.

Please note that the term "energy" refers to the energy content flowing across the system boundary. If, for example, the backup heating is a fuel-driven boiler, then the measurement point (and the energy flow at that point) for the fuel input has to be defined.

The representation of the proposal is shown in Figure 4.3 and the nomenclature and abbreviations used are provided in the Nomenclature of the book. The representation method used is based on the work by Frank et al. [6]. However, as the system in Figure 4.3 represents a generic configuration, some simplifications have been made in order not to overload the representation:

- Ambient and exhaust air, groundwater, ground, and waste heat can all be considered as heat sources for the heat pump ($\dot{Q}_{HS}$) or heat sinks in the case of free cooling or ground regeneration with excess solar heat ($\dot{Q}_{FC}$) or energy dissipation for active cooling ($\dot{Q}_{HR}$). They, as well as the respective heat exchangers, were put together as "free energy sources" and indicated with an orange frame (slashed for heat exchangers).
- Solar collectors can generally transform both solar radiation and heat from ambient air (including latent heat) into useful heat or heat source for the heat pump (either directly or for the regeneration of the ground, air preheating, etc.). This fact has been considered by putting air and sun together within the yellow frame. The energy input to the collector is denominated as $G \cdot A_{coll}$ (which equals the total irradiation on the collectors) for the solar radiative part and $\dot{Q}_{coll,air}$ for the energy input from the ambient air.
- Traded energy includes electricity and other energy carriers, denominated with "energy carrier X".

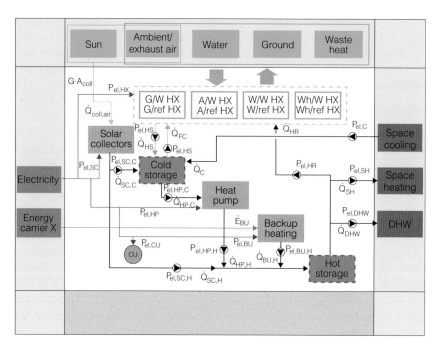

Fig. 4.3 Reference SHP system

- Component denominated as "CU" represents the electricity consumption of all control unit(s) of the system not included in the consumption of component-specific controls, for example, heat pump or solar system control. In many cases, however, it is very difficult or even impossible to clearly define the electricity consumption of the control unit(s) for some subsystems within their respective system boundaries. As always, common sense should be used and the chosen approach described.
- Obviously, the pumps and some other components not represented in Figure 4.3 (e.g., valves) also consume electricity. In order not to overload the representation, electricity consumption of the pumps is considered with the pump symbol and other components are not included. In real systems, however, their energy consumption should be considered appropriately.

In many systems, hydraulic connections allow energy transport to and from the heat pump bypassing the heat storage(s) or the system does not contain one or both of the represented heat storages. To avoid overloading the diagram with the representation of these energy flows, both storages have been represented within a dashed frame. This means that in case a storage is bypassed or not existent in a system, it should not be taken into account or it represents just a hydraulic connection. For example, if the solar collector and a borehole are both directly connected to the evaporator of the heat pump, then the cold storage is reduced to a nodal point for $\dot{Q}_{HS}$, $\dot{Q}_{SC,C}$, and $\dot{Q}_{HP,C}$.

4.3 Reference system and system boundaries

Note that the energy flows are represented in their physical flow direction, from higher to lower temperatures. The connections between the components do not necessarily reproduce the hydraulic configuration of the system. They, however, provide information on possible interactions between the components, due to the hydraulics and the control strategy of the system. The connections between components with a pump symbol represent energy consumption needed to transport the heat transfer medium and overcome the pressure losses within the system. These consumptions generally represent a substantial part of the system energy input and have to be considered appropriately. In some systems, one pump can be used to transport the heat transfer medium from one to several components. For example, one pump can be used to circulate the fluid from the collector both to the evaporator of the heat pump and to the heat storage. This implies that this pump would be consuming both $P_{el,SC,H}$ and $P_{el,SC,C}$. This has to be considered when balancing the system using measurement data or simulation results. In analogy, one connection from Figure 4.3 can in reality contain more than one circulation pump due to heat exchangers. For example, although presented as one component, "hot storage" can actually consist of more than one unit (e.g., one storage for heating and one for DHW). This implies that, for example, the energy input $P_{el,SC,H}$ can in reality consist of more than one consumer (pump). This has to be taken into account when calculating the overall energy input to the system.

4.3.2 Definition of system boundaries and corresponding seasonal performance factors

As previously mentioned, the aim of the definition of system boundaries was to establish generally applicable principles, largely regardless of the technology and system configuration. The principles were therefore discussed and proposed together with another international activity, IEA HPP Annex 34 – Thermally Driven Heat Pumps for Heating and Cooling (www.annex34.org), which focused on thermally driven heat pumps for heating and cooling. Similar approach developed within the project SEPEMO-Build (Seasonal Performance Factor and Monitoring for Heat Pump Systems in the Building Sector, www.sepemo.eu) for electrically driven compression heat pump systems in buildings was also considered. When defining the boundaries, two main goals were pursued:

- The boundaries should allow a simple application not only to different SHP systems, but also to other technologies for transparent comparison of different products regarding different performance aspects (energy, economics, environment, etc.).
- The boundaries should cover the needs of different target groups and allow for a simple analysis of a system operation by comparing the performance within different system boundaries.

Starting from these two goals, the following five main principles for the definition of system boundaries were proposed in Table 4.1. When applied to the reference system from Figure 4.3, specific system boundaries for SHP systems can be defined as

1. SHP system with the useful energy distribution systems (SHP+);
2. SHP system without the useful energy distribution systems (SHP);
3. SHP system without the useful energy storage (bSt);
4. heat pump with the heat source/heat rejection subsystems (HP + HS/HP + HR);
5. heat pump, solar collector, and backup unit (HP, SC, and BU).

Table 4.1 Overview of main principles for the definition of system boundaries for performance evaluation of SHP systems

System boundary	Purpose	Target group
Overall system performance including energy distribution system	Possibility of an energy-, economy-, and ecology-related evaluation of the whole system – overall energy balance, traded energy, free energy, emissions, and so on	Users, policy makers, statistical evaluation
Overall system performance without the energy distribution system	Possibility of an energy-, economy-, and ecology-related evaluation of the energy producing system, without the energy distribution system, which may vary for different applications. Comparison between different systems and technologies, product quality assurance, and labeling	Manufacturers, planners, installers, users, funding institutions, policy makers
Performance of the system without the influence of the end user storage losses	Mainly interesting for system analysis – storage management	System and component manufacturers, planners
Performance of each energy transformation unit, including all parts needed for its proper functioning	Performance of each unit under given circumstances gives information about the efficiency of every subsystem and possible improvements	Component and subcomponent manufacturers, planners, installers
Performance of each energy transformation unit itself, without influence of the auxiliary energy	This closely corresponds to the energy balance used currently in most quality assurance schemes for both solar thermal collectors and heat pumps (e.g., Solar Keymark, EHPA Quality Label). By comparison with other performance figures, an analysis of the system regarding peripheral energy consumption can be made	System and component manufacturers, planners, installers

4.3 Reference system and system boundaries

The applications of the principles from Table 4.1 on SHP reference system for heating and cooling applications are shown in Figures 4.4 and 4.6–4.8. Note that boundaries bSt and HP + HS (HP + HR) differ for heating and cooling operation modes. In the first case, this is due to the definition of the useful energy storage. In the heating operation mode, the useful energy is stored in the hot storage, while the cold storage (if any) can be part of the energy source system for the heat pump. In the cooling operation mode, the energy extracted from the user, that is, the useful cooling energy, is stored in the cold storage and this component is excluded from the energy balance. For the HP + HS and HP + HR subsystems, the heat source for the heating and heat sink system for the cooling operation mode differ and respective components have to be considered accordingly. Further, due to the complexity of the schematic, the convention regarding balancing of all energy flows crossing the system boundary cannot in all cases be fully complied with. For example, $\dot{Q}_{HR}$ from Figure 4.6a is crossing the system boundary, but is not considered in the balance. In such cases, compare the figure with the respective equation defining a performance figure for better understanding.

Electricity consumption for liquid pumps between the system components is included in the energy balance only if both components are considered within the boundary. In some cases, this might lead to unrealistic performance figures, as described for the system boundary bSt. As always, common sense has to be used for the interpretation of calculated values in such cases.

Boundaries SHP+ and SHP (Figure 4.4) are recommended for the comparison of SHP systems among each other, as well as for the assessment of the environmental impact of the systems in operation. Depending on the available data or purpose of the comparison, the one or the other boundary might be more appropriate.

The useful energies for space heating and cooling within both system boundaries are considered at the interface to the energy distribution system, for example, at the heating circuit manifold. However, for the SHP+ boundary the energy input of the distribution system for circulation pumps, ventilators, controls, and so on is considered and for the SHP boundary it is excluded from the energy balance.

SPF_{SHP+} and SPF_{SHP} can be calculated according to Equations 4.15 and 4.16, respectively. The given equations consider heating, cooling, and DHW preparation altogether as useful energy delivered to the customer. In most cases, however, the performance of the system while delivering energy for space heating or space cooling only or space heating and domestic hot water is of interest. In these cases, only the electricity consumption of the system for that (these) operation modes(s) has to be considered. In many cases, this is a very straightforward process, but in more complex systems some assumptions have to be made. For more information on some difficulties applying this method to systems delivering more than one useful energy at time, see the textbox in Section 4.3.2.

$$SPF_{SHP+} = \frac{\int (\dot{Q}_{SH} + \dot{Q}_{DHW} + \dot{Q}_{C})dt}{\int (\sum P_{el,SHP+})dt},$$

$$\sum P_{el,SHP+} = P_{el,SC} + P_{el,SC,C} + P_{el,SC,H} + P_{el,HP} + P_{el,HP,C} + P_{el,HP,H} + P_{el,HS} + P_{el,BU}$$
$$+ P_{el,BU,H} + P_{el,SH} + P_{el,DHW} + P_{el,C} + P_{el,FC} + P_{el,HR} + P_{el,HX} + P_{el,CU}.$$

(4.15)

80   4 Performance and its assessment

(a)

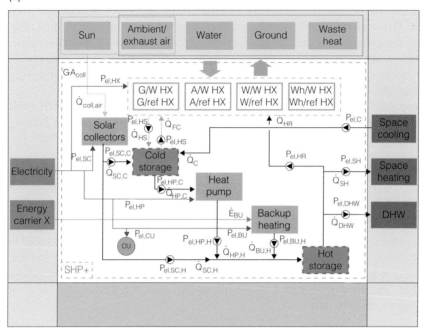

(b)

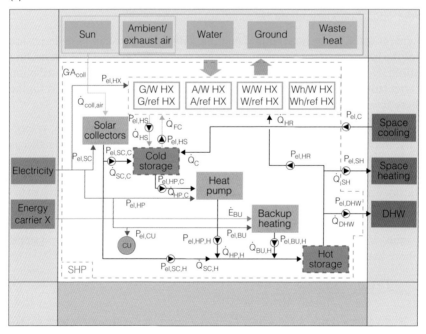

Fig. 4.4 Main system boundaries for the reference SHP system for heating and cooling applications: SHP+ system boundary (a) and SHP system boundary (b)

4.3 Reference system and system boundaries

Table 4.2 Consideration of DHW circulation system energy consumption for SHP and SHP+ system boundaries

	Direct system		Fresh water system	
	SHP	SHP+	SHP	SHP+
Useful energy	$\dot{Q}_{DHW}$	$\dot{Q}_{DHW}$	$\dot{Q}_{DHW}$	$\dot{Q}_{DHW}$
Energy consumption	—	$P_{el,DHW,circ}$	$P_{el,DHW,prim}$	$P_{el,DHW,prim} + P_{el,DHW,circ}$

$$SPF_{SHP} = \frac{\int (\dot{Q}_{SH} + \dot{Q}_{DHW} + \dot{Q}_{C}) dt}{\int (\sum P_{el,SHP}) dt},$$

$$\sum P_{el,SHP} = P_{el,SC} + P_{el,SC,C} + P_{el,SC,H} + P_{el,HP} + P_{el,HP,C} + P_{el,HP,H} + P_{el,HS} + P_{el,BU}$$

$$+ P_{el,BU,H} + P_{el,DHW} + P_{el,FC} + P_{el,HR} + P_{el,HX} + P_{el,CU}.$$

(4.16)

Note that Table 4.2 has to be considered for the assessment of electricity consumption for the distribution of DHW – $P_{el,DHW}$.

The delivered useful energy for DHW preparation taken into account generally depends on the system configuration:

a) *Systems without additional heating of the distribution pipes or secondary flow circulation*:
   - For fresh water systems (systems with a DHW heat exchanger – Figure 4.5b, however, without the circulation pump), the energy provided to the DHW distribution system at the secondary DHW lines, without the distribution losses after the connection (seen from the system point of view – after the heat exchanger).
   - For direct systems (Figure 4.5a, however, without the circulation pump), the energy provided to the DHW distribution system at the connection of the supply line at the storage or at the heat pump or any other heat generating unit delivering DHW as useful energy, that is, without the distribution losses after the connection.

b) *Systems with additional heating of the distribution pipes*:

In this case, the useful energy considered is the energy provided after the last energy input to the supply line, that is, before or at the tap. If a measurement before the tap is not convenient or not possible, an approximation can be made by adding the electricity consumption for the heating of the distribution pipes to both the overall consumed electricity (full amount) and useful energy output (fully, if heating is needed and it can be assumed that most of the heat is not dissipated to the environment; otherwise, a certain fraction has to be assumed depending on the controls, tapping cycle, building type, etc.).

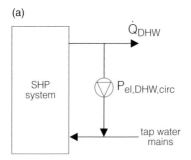

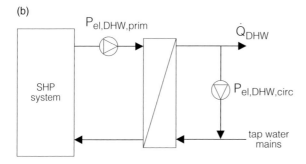

Fig. 4.5 Consideration of DHW secondary flow circulation for the energy consumption of the system regarding SHP and SHP+ system boundaries for direct systems (a) and fresh water systems (b)

c) *Systems with flow circulation*:

The useful energy is the same for both system configurations (direct and fresh water) and for both system boundaries – Q_{DHW}. However, the considered energy consumption attributed to the DHW distribution system varies for different system configurations, as shown in Figure 4.5 and Table 4.2.
- For *fresh water systems*, the energy consumption of the primary pump for domestic hot water preparation ($P_{el,DHW,prim}$) is included in both SHP and SHP+ boundaries, since it is considered to be an integral part of the SHP system. However, for system boundary SHP, only the energy consumption when delivering DHW is taken into account. Energy consumption of the circulation pump ($P_{el,DHW,circ}$) is considered only for the SHP+ system boundary.
- For *direct systems*, the energy consumption of the circulation pump ($P_{el,DHW,circ}$) is considered only for the SHP+ system boundary.

In both system boundaries, the heat losses after the circulation loop should not be regarded as useful energy Q_{DHW}.

Reverse heat flows from the user, not intended for space cooling, for example, for defrosting of air source heat pumps, have to be taken into account and subtracted from the useful energy provided.

4.3 Reference system and system boundaries

> **On difficulties regarding simultaneous operation or configurations including, for example, desuperheater**
>
> In some cases, it is very difficult or even impossible to attribute energy consumption of a system component to only one operation mode. For example, for heat pump units with a desuperheater for DHW production, there is no generally accepted method to separate the electricity consumption for DHW and heating operation modes when the unit is providing both energies simultaneously. The same applies for simultaneous heating and cooling or space cooling and DHW production. But even a simpler example such as free cooling operation when the heat from the building is extracted by circulating heat transfer media between a ground source and the building can prove to be hard to handle. In this case, it can be argued that while cooling the building, the ambient heat is being transferred to the heat source, thus attributing to better performance during heating operation. Again, how to split the electricity consumption of the liquid pump between the heating and the cooling operation mode? In these and some other cases, some assumptions using best engineering practice or calculations including second law analysis have to be made.

In the system boundaries bSt – before storage (Figure 4.6), a distinction between the heating operation mode (HOM) and the cooling operation mode (COM) is made, as previously discussed. All useful energy outputs of the SHP system to the energy storages are taken into account. Thus, the storage losses as well as the energy needed to supply the heat to the storages are not included. However, depending on the operation mode of the system, the function of the storages and the driving energy input to the system might vary. For example, in the heating operation mode, the cold storage is included in the system, since it is a part of the heat source for the heat pump. If switched to the cooling operation mode, the cooling load is supplied from the cold storage, which then has to be excluded from the energy balance. Hence, for systems capable of operating in several modes over the considered measurement or simulation period, the system boundaries have to change accordingly. This is not a substantial limitation, since most of the SHP systems described within T44A38 are designed primarily or solely for these applications. Nevertheless, the possibility to consider the cooling operation mode is also given.

The $SPF_{bSt,HOM}$ for the heating operation mode and $SPF_{bSt,COM}$ for the cooling operation mode are defined as

$$SPF_{bSt,HOM} = \frac{\int (\dot{Q}_{SC,H} + \dot{Q}_{HP,H} + \dot{Q}_{BU,H})dt}{\int (\sum P_{el,bSt,HOM})dt},$$

$$\sum P_{el,bSt,HOM} = P_{el,SC} + P_{el,SC,C} + P_{el,HP} + P_{el,HP,C} + P_{el,HS} + P_{el,FC} + P_{el,BU} + P_{el,HX} + P_{el,CU},$$

(4.17)

(a)

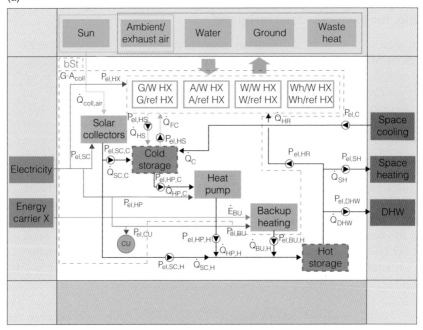

(b)

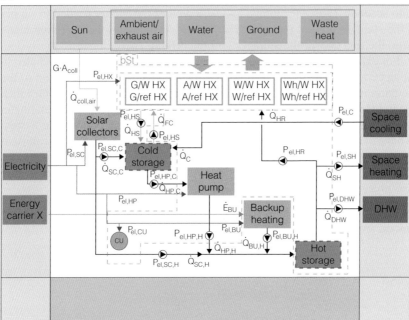

Fig. 4.6 Main system boundaries for the reference SHP system for heating and cooling applications: bSt system boundary for the heating operation mode (a) and bSt system boundary for the cooling operation mode (b)

4.3 Reference system and system boundaries

$$\text{SPF}_{\text{bSt,COM}} = \frac{\int \left(\dot{Q}_{\text{HP,C}} + \dot{Q}_{\text{FC}}\right) dt}{\int \left(\sum P_{\text{el,bSt,COM}}\right) dt}, \quad (4.18)$$

$$\sum P_{\text{el,bSt,COM}} = P_{\text{el,HP}} + P_{\text{el,HP,H}} + P_{\text{el,HR}} + P_{\text{el,HX}} + P_{\text{el,CU}}.$$

Due to the definition of the system boundaries, in some cases SPF_{bSt} will not be realistic. For example, if the system delivers useful cooling energy only by free cooling, the consumed electricity will consider only the control unit and the performance factor will be unproportional compared with performance factors for other system boundaries. Reporting such values should be avoided or adequately explained if reported.

Within the system boundaries HP + HS (HP + HR) – heat pump with heat source (heat pump with heat rejection) (Figure 4.7), the subsystem including heat pump unit with all its heat sources or heat sinks, which can also include another heat transforming unit (here the solar collector), is balanced. As for the previous system boundaries (bSt), different operation modes have to be considered separately because of different resulting boundaries. In the space heating and DHW operation mode, the useful energy output considered is the gross energy output of the heat pump only, not taking into account losses due to short- or long-term storage, piping, and so on. The energy input to the system includes the overall input needed for both the heat pump and the considered solar thermal parts. If the solar collectors do not interact with the heat pump directly (e.g., direct evaporation in the collector) or indirectly (e.g., feeding into the heat source of the heat pump – ground, low-temperature storage, etc.), then they should be excluded from the calculations. The overall system control unit is not included in the energy balance, since only the heat pump system is considered. However, if the heat pump does not have an additional control unit, the consumption of the central one should be included appropriately.

$\text{SPF}_{\text{HP+HS}}$ for the heating operation mode and $\text{SPF}_{\text{HP+HR}}$ for the cooling operation mode can be calculated as

$$\text{SPF}_{\text{HP+HS}} = \frac{\int \dot{Q}_{\text{HP,H}} \, dt}{\int \left(P_{\text{el,HP}} + P_{\text{el,HP,C}} + P_{\text{el,SC}} + P_{\text{el,SC,C}} + P_{\text{el,HS}} + P_{\text{el,HX}} + P_{\text{el,FC}}\right) dt}, \quad (4.19)$$

$$\text{SPF}_{\text{HP+HR}} = \frac{\int \dot{Q}_{\text{HP,C}} \, dt}{\int \left(P_{\text{el,HP}} + P_{\text{el,HP,H}} + P_{\text{el,HR}} + P_{\text{el,HX}}\right) dt}. \quad (4.20)$$

In the system boundary HP – heat pump (Figure 4.8), the consumed electricity includes supporting systems such as controls and crankcase heaters for the entire measurement or simulation period, including standby periods. It is similar to the one currently used for the definition of the COP and SCOP in a number of European standards. There, the energy inputs and outputs are often corrected as described in, for example, EN 14511 (see Section 4.2.4.1). If used for a performance figure calculated from the field trial results, the energy output will generally not be corrected due to considerable effort needed to obtain all measurement data needed.

(a)

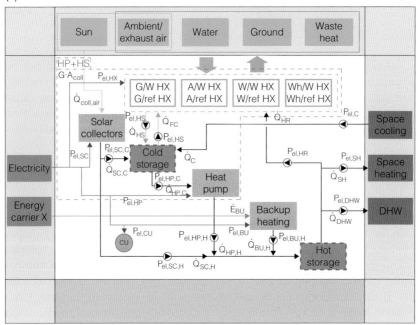

(b)

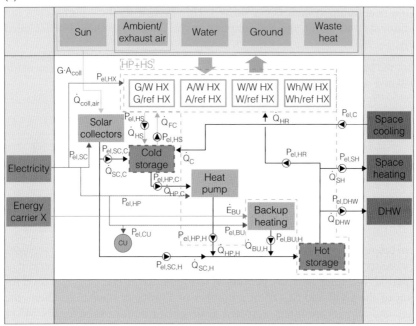

Fig. 4.7 Main system boundaries for the reference SHP system for heating and cooling applications: HP + HS system boundary (a) and HP + HR system boundary (b)

4.4 Environmental evaluation of SHP systems

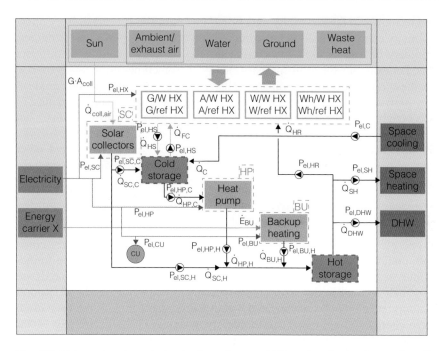

Fig. 4.8 Main system boundaries for the reference SHP system for heating and cooling applications: SC, HP, and BU system boundaries

SPF$_{HP}$ values for the heating and the cooling operation modes are given in Equations 4.21 and 4.22, respectively:

$$\text{SPF}_{\text{HP,HOM}} = \frac{\int \dot{Q}_{\text{HP,H}} \, dt}{\int P_{\text{el,HP,HOM}} \, dt}, \tag{4.21}$$

$$\text{SPF}_{\text{HP,COM}} = \frac{\int \dot{Q}_{\text{HP,C}} \, dt}{\int P_{\text{el,HP,COM}} \, dt}. \tag{4.22}$$

The performance figures for the backup unit (BU) will generally depend on the technology of the applied device. Figures for solar collector (SC) efficiency can be found in standards and guidelines and are briefly discussed in Section 4.4. Note that $P_{\text{el,SC}}$ considers only direct energy consumption of the solar collector like a fan for hybrid collectors. For standard solar collectors, SC is not defined as the energy consumption for the circulation pump $P_{\text{el,SC,H}}$ is not considered.

4.4 Environmental evaluation of SHP systems

For a transparent energy balancing of an SHP system, or any energy transformation unit, and especially for the evaluation of its impact on the environment due to the usage of locally available energy sources, knowledge about the energy supply chain from the used natural resources to the electricity plug or gas meter on site is needed. Very often, the term "primary energy" is used in this context. However, as for performance figures, the term primary energy is used to describe different energy qualities in different publications. For example, the term "primary energy" is sometimes erroneously used to denominate the energy input that is available at the interface to the energy using product, for example, SHP system, such as electricity at the plug, natural gas from the mains, or wood pellets bought at the local store. However, these energy sources are themselves but a product from other natural resources down the supply chain. For their production, transportation, and distribution to the site of consumption, more energy and other resources had to be used and losses due to transportation and conversion processes have to be accounted for.

Therefore, a clear definition of different energy types is needed. Taking definitions provided in Refs [7,8], the following types of energies are defined for this book:

- *Primary energy:* Energy contained in various natural resources, unprocessed – crude oil, wood, uranium, wind, or solar energy. Primary energy sources can be further divided into "renewable" and "nonrenewable" sources. Nonrenewable sources are limited by their amount in the accessible environment (e.g., coal, crude oil, natural gas, uranium) and cannot be regenerated within a reasonable time frame once used. Renewable energy sources either are limitlessly available in the nature (e.g., solar irradiation and tidal or wind energy) or can be regenerated on the human timescale (biomass).
- *Final energy:* Energy supplied to the user. This type of energy has generally undergone some kind of conversion – refining, enrichment, and purification – and has been transported to the site of consumption/purchase. Examples for this energy type are electricity or gas at the meter, wood pellets, or electricity from a PV panel.
- *Useful energy:* Energy provided to the user at its final form for the intended purpose – heating, cooling, transportation, leisure, and entertainment – and is largely dissipated in the course of the consumption.

Figure 4.9 shows the differences between these three energy types.

Note that the final energy coming from renewable resources (electricity from PV and thermal energy from biomass or solar panels) is not 100% renewable due to production,

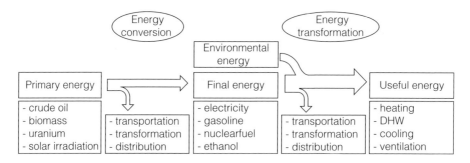

Fig. 4.9 Difference between primary energy, final energy, and useful energy

4.4 Environmental evaluation of SHP systems

transportation, and installation processes, among others, which generally require usage of nonrenewable resources, including energy. This so-called gray energy is being only partially considered in the performance figures defined subsequently, through CED_{NRE} and GWP_{ec} factors.

In order to compare systems and technologies in terms of their environmental impact, two main performance figures are recommended in this chapter:

- Primary energy ratio of nonrenewable energy sources – PER_{NRE}.
- Equivalent warming impact of the SHP system – EWI_{sys}.

PER_{NRE} gives information on the consumption of nonrenewable energy sources for the provision of useful energy output of the system. Note that it does not account for the production, distribution, installation, and end-of-life disposal of the SHP unit or system itself. It is a figure that considers the usage of limited energy resources contained in, for example, fossil fuels. It is defined as the ratio of the useful energy output of the system to the primary energy input.

EWI_{sys} is the ratio of the greenhouse gas emission to the useful energy output of the system. The greenhouse gas emission is expressed as equivalent CO_2 emission for the provision of the energy carrier on installation site (final energy) and its consumption.

For the calculation of these two figures, the following indicators are needed:

- CED_{NRE} – *cumulative energy demand (CED), nonrenewable:* It quantifies the nonrenewable primary energy used to provide the final energy, including the energy used for construction of the electric grid and power plants. This indicator accounts for the primary energy from fossil, nuclear, and primary forest resources (i.e., original forests that are destroyed and replaced by farmland) defined in terms of primary energy to final energy – kWh_{pe}/kWh_{fe}.
- GWP_{ec} – *global warming potential:* It is the weighted addition of the emission of different greenhouse gases when providing final energy, including emissions generated during construction of the electric grid and power plants. It does not take into account refrigerant leakage during the system operation. It is expressed in terms of equivalent quantity of carbon dioxide per quantity of final energy (kg CO_2 equiv./ kWh_{fe}) for a time frame of 100 years. Please note that the GWP_{ec} as defined here differs from the GWP used to quantify the influence of the various substances when released into the atmosphere on the global warming phenomena (as defined, for example, in Ref. [9]).

Since the provenance of the electrical energy at the plug varies widely from country to country due to their power generation and import mixes, it is important to define reference values for comparison purposes. For the electrical energy, the corresponding European electricity supply mix (ENTSO-E– European Network of Transmission System Operators for Electricity) on low voltage level for these two indicators was chosen from Ref. [10]. They include electricity production, transmission, and distribution including corresponding losses. In certain cases, however, it is favorable to use specific national values, which may differ substantially between the countries. These values can be found in, for example, Ref. [10]. For transnational comparisons, values from Table 4.3 are recommended.

Table 4.3 CED_{NRE} and GWP_{ec} for different energy carriers

Energy carrier	CED_{NRE} (kWh$_{pe}$/kWh$_{fe}$)	GWP_{ec} (kg CO_2 equiv./kWh$_{fe}$)
Electricity	2.878	0.521
Gas	1.194	0.307
Oil	1.271	0.318
Wood		
Logs	0.030	0.020
Pellets	0.187	0.041
Chips	0.035	0.011

Sources: Refs [10,11].

For all other energy carriers, the values for each country are nearly identical and are taken from the Ecoinvent database [11] that contains a large number of processes for production of goods and provision of services with a focus on European production chains (see Table 4.3).

Subsequently, these values will be used to define the primary energy ratio, nonrenewable (PER_{NRE}) and the EWI_{sys} factor for each investigated system.

4.4.1.1 Primary energy ratio

To relate the useful energy output to the nonrenewable primary energy consumption, the primary energy ratio (PER_{NRE}) is defined with the unit kWh$_{ue}$/kWh$_{pe}$.

For an electric system, it is calculated as

$$PER_{NRE} = \frac{\int (\dot{Q}_{SH} + \dot{Q}_{DHW} + \dot{Q}_C) dt}{\int \sum (P_{el,final} \cdot CED_{NRE,el}) dt}, \tag{4.23}$$

where $P_{el,final}$ is the total electricity consumption of the system during operation.

By introducing the definition of the SPF (Equation 4.9), it can be transformed to

$$PER_{NRE} = \frac{SPF_{SHP}}{CED_{NRE,el}}. \tag{4.24}$$

For systems using different primary energy sources, this factor is calculated as

$$PER_{NRE} = \frac{\int (\dot{Q}_{SH} + \dot{Q}_{DHW} + \dot{Q}_C) dt}{\int \sum_{i=\text{energy source}} (\dot{Q}_{fe,i} \cdot CED_{NRE,i}) dt}, \tag{4.25}$$

where $Q_{fe,i}$ is the final energy consumption of the system during operation and expressed in kWh$_{fe}$.

4.5 Calculation example

In some cases, the reciprocal value of PER_{NRE} – primary energy effort figure (PEEF) can be useful:

$$PEEF_{NRE} = \frac{\int \sum_{i=\text{energy source}} (\dot{Q}_{fe,i} \cdot CED_{NRE,i}) dt}{\int (\dot{Q}_{SH} + \dot{Q}_{DHW} + \dot{Q}_C) dt} = \frac{1}{PER_{NRE}}. \qquad (4.26)$$

4.4.1.2 Equivalent warming impact

Similarly to Equation 4.24, but with SPF in denominator, the EWI_{sys} figure (kg CO_2 equiv./kWh$_{ue}$) for an electric system can be defined as

$$EWI_{sys} = \frac{GWP_{el}}{SPF_{SHP}}. \qquad (4.27)$$

For systems using other primary energy sources, this results in

$$EWI_{sys} = \frac{\int \sum_{i=\text{energy source}} (\dot{Q}_{fe,i} \cdot GWP_{ec,i}) dt}{\int (\dot{Q}_{SH} + \dot{Q}_{DHW} + \dot{Q}_C) dt}. \qquad (4.28)$$

4.4.1.3 Fractional primary energy savings

Using the formulations from the previous section, fractional energy savings can be defined for the SHP system. If there is more than one energy carrier in the system, the fractional energy savings have to be related to the primary energy. All final energy consumptions have to be multiplied with the primary energy factors (Equation 4.29). Furthermore, it is also important to specify the system boundary (here SHP).

$$f_{sav,SHP,pe} = 1 - \frac{\left(\sum_{i=\text{energy source}} (\dot{Q}_{fe,i} \cdot CED_{NRE,i})\right)_{SHP}}{\left(\sum_{i=\text{energy source}} (\dot{Q}_{fe,i} \cdot CED_{NRE,i})\right)_{ref}}. \qquad (4.29)$$

4.4.1.4 Fractional CO$_2$ emission savings

Equivalently to Section 4.4.1.3, the fractional CO_2 emission savings can be defined (Equation 4.30). These are balanced according to the system boundary SHP.

$$f_{sav,SHP,emission} = 1 - \frac{\left(\sum_{i=\text{energy source}} (\dot{Q}_{fe,i} \cdot GWP_{ec,i})\right)_{SHP}}{\left(\sum_{i=\text{energy source}} (\dot{Q}_{fe,i} \cdot GWP_{ec,i})\right)_{ref}}. \qquad (4.30)$$

4.5 Calculation example

For better understanding of the application of system boundary definitions to real systems and the meaning of different performance figures, a calculation example is given.

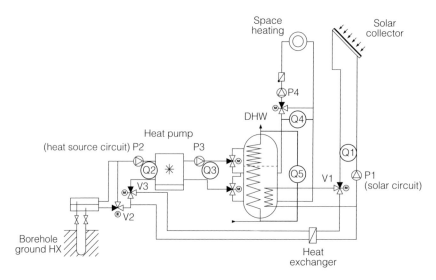

Fig. 4.10 Simplified hydraulic representation of a real SHP system

In Figure 4.10, an existing SHP system is represented with a simplified hydraulic schematic. The same system can be easily represented with the energy flow diagram developed in T44A38. System boundaries from Section 4.3 are applied.

Comparing Figures 4.10 and 4.11, the following can be observed:

– As there is no cold storage in the system, the component "cold storage" from the square view representation is reduced to a nodal point that represents the heat exchanger between the solar circuit and heat source circuit from Figure 4.10.
– Electricity consumption $W_{el,1}$ of pump P1 from Figure 4.10 is represented by the electricity consumptions $P_{el,SC,C}$ and $P_{el,SC,H}$ in Figure 4.11. Hence, for certain performance figures, the knowledge of the position of magnetic valves V1, V2, and V3 has to be known in order to correctly attribute the electricity consumption of P1 during different operation modes of the system, that is, charging of the storage from the solar circuit, regeneration of the ground with solar heat over the heat source circuit, or direct usage of heat from the solar circuit via heat source circuit at the evaporator of the heat pump.
– Electricity consumption $W_{el,2}$ of pump P2 from Figure 4.10 is represented by the electricity consumptions of $P_{el,HS}$, $P_{el,HP,C}$, and $P_{el,FC}$ (Figure 4.11). The same considerations as for pump P1 apply.
– As it is assumed that only a minimum of heat flows will be measured (heat meters Q1–Q5 in Figure 4.10), for a more in-depth evaluation of the system performance according to the proposed system boundaries from the measurement data, time-resolved data sets for most of the heat flows, electricity consumptions, and valve positions are needed.

4.5 Calculation example

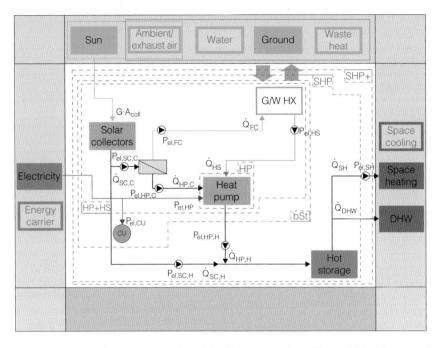

Fig. 4.11 Energy flow representation of the SHP system from Figure 4.10 with system boundaries according to Chapter 4: calculation example

- The consumption of magnetic valves has also to be taken into consideration, if possible.
- DHW system is a direct system with neither circulation nor additional heating of distribution pipes.

From the measurement, the following values are known for a period of 1 year. In addition, time-resolved values in 1 min steps, including positions of all valves, are available.

Q_1	6000 kWh
Q_2	7200 kWh
Q_3	10 200 kWh
Q_4	11 500 kWh
Q_5	2700 kWh

$W_{el,1}$	450 kWh
$W_{el,2}$	550 kWh
$W_{el,3}$	350 kWh
$W_{el,4}$	400 kWh
$W_{el,HP}$	2000 kWh
$W_{el,CU}$	175 kWh

From Figures 4.10 and 4.11 and knowing the operation modes of the system, following correlations can be derived:

$Q_1 = Q_{SC,H} + Q_{SC,C}$
$Q_2 = Q_{HP,C} + Q_{FC} + Q_{HS}$
$Q_3 = Q_{HP,H}$
$Q_4 = Q_{SH}$
$Q_5 = Q_{DHW}$

$W_{el,1} = W_{el,SC,H} + W_{el,SC,C}$
$W_{el,2} = W_{el,HS} + W_{el,HP,C} + W_{el,FC}$
$W_{el,3} = W_{el,HP,H}$
$W_{el,4} = W_{el,SH}$

With time-resolved measurement data, further correlations can be made:

$Q_1 = Q_{SC,C}$	$W_{el,1} = W_{el,SC,C}$	when V1 is directing the flow of the solar circuit to the heat exchanger
$Q_1 = Q_{SC,H}$	$W_{el,1} = W_{el,SC,H}$	when V1 is directing the flow of the solar circuit to the heat storage
$Q_2 = Q_{HP,C}$	$W_{el,2} = W_{el,HS}$	when V1 is directing the flow of the solar circuit to the heat exchanger and V2 and V3 are directing the flow of the heat source circuit to the evaporator
$Q_2 = Q_{FC}$	$W_{el,2} = W_{el,HP,C}$	when V1 is directing the flow of the solar circuit to the heat exchanger and V2 and V3 are directing the flow of the heat source circuit to the borehole
$Q_2 = Q_{HS}$	$W_{el,2} = W_{el,FC}$	when V2 and V3 are directing the flow of the heat source circuit from the borehole to the evaporator

Thus, from the measurement data and using the above correlations the following values for heat flows and electricity consumptions according to Figure 4.11 are obtained:

Q_{FC}	700 kWh
$Q_{HP,C}$	500 kWh
$Q_{SC,C}$	1200 kWh
Q_{HS}	6700 kWh
$Q_{SC,H}$	4800 kWh
$Q_{HP,H}$	10 200 kWh
Q_{SH}	11 500 kWh
Q_{DHW}	2700 kWh

$W_{el,FC}$	150 kWh
$W_{el,HP,C}$	100 kWh
$W_{el,SC,C}$ [a]	250 kWh
$W_{el,HS}$	300 kWh
$W_{el,SC,H}$	200 kWh
$W_{el,HP,H}$	350 kWh
$W_{el,SH}$	400 kWh
$W_{el,HP}$	2000 kWh
$W_{el,CU}$	175 kWh

[a] Sum of $W_{el,FC} + W_{el,HP,C}$.

4.5 Calculation example

Using Equations 4.15–4.22, seasonal performance figures can be calculated as follows:

$$\text{SPF}_{\text{SHP+}} = \frac{Q_{\text{SH}} + Q_{\text{DHW}}}{\sum W_{\text{el,SHP+}}} = \frac{14\,200}{3675} = 3.86,$$

$$\sum W_{\text{el,SHP+}} = W_{\text{el,SC,C}} + W_{\text{el,SC,H}} + W_{\text{el,HP}} + W_{\text{el,HS}} + W_{\text{el,HP,H}} + W_{\text{el,SH}} + W_{\text{el,CU}}$$
$$= 3675 \text{ kWh},$$

$$\text{SPF}_{\text{SHP}} = \frac{Q_{\text{SH}} + Q_{\text{DHW}}}{\sum W_{\text{el,SHP}}} = \frac{14\,200}{3275} = 4.34,$$

$$\sum W_{\text{el,SHP}} = W_{\text{el,SC,C}} + W_{\text{el,SC,H}} + W_{\text{el,HP}} + W_{\text{el,HS}} + W_{\text{el,HP,H}} + W_{\text{el,CU}} = 3275 \text{ kWh},$$

$$\text{SPF}_{\text{bSt}} = \frac{Q_{\text{HP,H}} + Q_{\text{SC,H}}}{\sum W_{\text{el,bSt}}} = \frac{15\,000}{2725} = 5.50,$$

$$\sum W_{\text{el,bSt}} = W_{\text{el,SC,C}} + W_{\text{el,HP}} + W_{\text{el,HS}} + W_{\text{el,CU}} = 2725 \text{ kWh},$$

$$\text{SPF}_{\text{HP+HS}} = \frac{Q_{\text{HP,H}}}{\sum W_{\text{el,HP+HS}}} = \frac{10\,200}{2550} = 4.00,$$

$$\sum W_{\text{el,HP+HS}} = W_{\text{el,SC,C}} + W_{\text{el,HP}} + W_{\text{el,HS}} = 2550 \text{ kWh},$$

$$\text{SPF}_{\text{HP}} = \frac{Q_{\text{HP,H}}}{W_{\text{el,HP}}} = \frac{10\,200}{2000} = 5.10.$$

Note that in this example also $P_{\text{el,FC}} + P_{\text{el,HP,C}}$ instead of $P_{\text{el,SC,C}}$ can be used, since both $P_{\text{el,FC}}$ and $P_{\text{el,HP,C}}$ are included in all system boundaries as $P_{\text{el,SC,C}}$. It seems worth mentioning again that a caution is advised when attributing measurement data from the real system to the components in the energy flow diagram!

In Figure 4.12, the seasonal performance factors for five boundaries are shown in a graph. Starting from the left, it can be seen that moving the system boundaries closer to the heat sources has a positive effect on the SPF by reducing the heat losses within and between the components, as well as the electricity consumption of the pumps. The gap

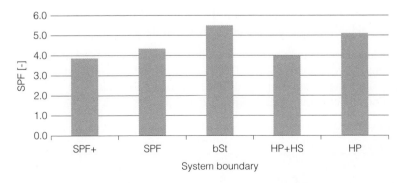

Fig. 4.12 Seasonal performance factors for five system boundaries defined within T44A38

between the boundaries bSt and HP + HS is caused by reducing the useful heat output only to the heat pump. An increase of SPF from HP + HS to HP is due to the elimination of the electricity consumption of the heat source of the heat pump.

For the analysis of the system, the following considerations regarding comparison of "adjacent" system boundaries might be helpful:

SPF+ → SPF: Influence of the useful energy distribution system.
SPF → bSt Influence of the losses in the heat storage and electricity consumption of, for example, liquid pumps needed for storage charging.
bSt → HP + HS Direct contribution of the solar system to charging the heat storage.
HP + HS → HP Influence of the heat source of the heat pump.

Furthermore, the solar fractions and the renewable energy fraction according to Equations 4.10–4.12 (Section 4.2.6) can be calculated as

$$f_{sol,1} = \frac{Q_{SC,H}}{Q_{DHW} + Q_{SH}} = \frac{4800}{14\,200} = 0.34,$$

$$f_{sol,2} = 1 - \frac{Q_{HP,H} + Q_{BU,H}}{Q_{DHW} + Q_{SH}} = 1 - \frac{10\,200}{14\,200} = 0.28,$$

$$f_{sol,3} = \frac{Q_{SC,H}}{Q_{HP,H} + Q_{SC,H} + Q_{BU,H}} = \frac{4800}{15\,000} = 0.32.$$

This means that, depending on the definition of the solar fraction, the percentage of solar energy in the useful energy mix ($f_{sol,1}$ and $f_{sol,2}$) or energy provided to the heat storage ($f_{sol,3}$) – excluding heat source regeneration and direct usage on the evaporator – varies between 28 and 34%.

$$f_{ren,SHP} = 1 - \frac{\int (\sum P_{el,SHP}) dt}{Q_{DHW} + Q_{SH}} \hat{=} 1 - \frac{1}{SPF_{SHP}} = 1 - \frac{1}{4.34} = 0.77.$$

From the calculation of the renewable energy fraction (Equation 4.13), it can be concluded that 77% of the useful energy, excluding used electricity, comes from renewable sources – solar collector and ambient heat from the ground. This figure is a pure energy balance of the SHP system not taking into account the provenience of the final energy.

With provided data and using Equations 4.25–4.27, the system can also be evaluated for its environmental impact. Note that PER_{NRE}, $PEEF_{NRE}$, and EWI_{sys} strongly depend on the chosen factors and may substantially differ for same SPF values at different installation sites. For this example, values from Table 4.3 were used.

$$PER_{NRE} = \frac{SPF_{SHP}}{CED_{NRE,el}} = \frac{4.34}{2.878} = 1.51,$$

$$PEEF_{NRE} = \frac{1}{PER_{NRE}} = \frac{1}{1.51} = 0.66.$$

PER$_{NRE}$ of 1.51 means that from one unit of nonrenewable primary energy used for the production of electricity that drives the system within its SHP system boundaries, 1.51 units of useful heat are delivered. For comparison, for direct electrical heating this ratio would be 1/2.878 or 0.35. PEEF$_{NRE}$ of 0.66 means that 0.66 units of nonrenewable primary energy are used to obtain 1 unit of delivered useful energy.

$$\text{EWI}_{sys} = \frac{\text{GWP}_{el}}{\text{SPF}_{SHP}} = \frac{0.521}{4.34} = 0.12.$$

The value of 0.12 for EWI$_{sys}$ means that the equivalent of 0.12 kg carbon dioxide was released for 1 kWh of useful energy delivered.

Finally, let us have a look at the primary energy and emission savings according to Sections 4.4.1.3 and 4.4.1.4 with Equations 4.29 and 4.30. For that calculation, a reference system has to be chosen. We will choose a gas boiler with an overall annual efficiency η_{boiler} of 0.9 or 90% for both space heating and DHW production. For measured annual useful heat delivery of 14 200 kWh, the gas consumption on site would be $14\,200 \times 0.9 = 18\,838.67$ kWh.

$$f_{sav,SHP,pe} = 1 - \frac{\sum W_{el,SHP} \cdot \text{CED}_{NRE,el}}{(Q_{SH} + Q_{DHW}) \cdot \eta_{boiler} \cdot \text{CED}_{NRE,gas}} = 1 - \frac{3275 \times 2.878}{14\,200 \times 0.9 \times 1.194} = 0.50,$$

$$f_{sav,SHP,emission} = 1 - \frac{\sum W_{el,SHP} \cdot \text{GWP}_{el}}{(Q_{SH} + Q_{DHW}) \cdot \eta_{boiler} \cdot \text{GWP}_{gas}} = 1 - \frac{3275 \times 0.521}{14\,200 \times 0.9 \times 0.307} = 0.57.$$

As shown, for the assumed SHP and reference systems, the SHP system would have 50% of the primary energy consumption and 57% of the equivalent CO_2 emissions of the reference gas boiler delivering the same amount of useful heat.

Appendix 4.A Reviewed standards and other normative documents

4.A.1 Heat pumps (Table 4.A.1)

CEN (2011) EN 14511:2011 – Air conditioners, liquid chilling packages and heat pumps with electrically driven compressors for space heating and cooling. CEN, Brussels, Belgium.

CEN (2011) EN 15879-1:2011 – Testing and rating of direct exchange ground coupled heat pumps with electrically driven compressors for space heating and/or cooling. Direct exchange-to-water heat pumps. CEN, Brussels, Belgium.

CEN (2011) EN 16147:2011 – Heat pumps with electrically driven compressors. Testing and requirements for marking for domestic hot water units. CEN, Brussels, Belgium.

AHRI (1998) AHRI Standard 320-98 – Water-Source Heat Pumps. AHRI, Arlington, USA.

AHR (1998) AHRI Standard 325-98 – Ground Water-Source Heat Pumps. AHRI, Arlington, USA.

AHRI (1998) AHRI Standard 330-98 – Ground Source Closed-Loop Heat Pumps. AHRI, Arlington, USA.

ISO (1998) ISO 13256-1:1998 – Water-source heat pumps. Testing and rating for performance. Part 1. Water-to-air and brine-to-air heat pumps. International Organization for Standardization, Geneva, Switzerland.

ISO (1998) ISO 13256-2:1998 – Water-source heat pumps. Testing and rating for performance. Part 2. Water-to-water and brine-to-water heat pumps. International Organization for Standardization, Geneva, Switzerland.

CEN (2011) EN 14825:2011 – Air conditioners, liquid chilling packages and heat pumps, with electrically driven compressors, for space heating and cooling. Testing and rating at part load conditions and calculation of seasonal performance. CEN, Brussels, Belgium.

ASHRAE (2010) ASHRAE 116-2010 – Methods of Testing for Rating Seasonal Efficiency of Unitary Air Conditioners and Heat Pumps. ASHRAE, Atlanta, USA.

VDI (2003) VDI 4650-1 – Calculation of heat pumps. Simplified method for the calculation of the seasonal performance factor of heat pumps. Electric heat pumps for space heating and domestic hot water. VDI, Düsseldorf, Germany.

CEN (2008) EN 15316-4-2:2008 – Heating systems in buildings. Method for calculation of system energy requirements and system efficiencies. Part 4-2. Space heating generation systems, heat pump systems. CEN, Brussels, Belgium.

Table 4.A.1 Overview of the performance figures for heat pumps and heat pump systems defined in different standards and guidelines

Heat pump standards		
Standard/guideline	PF	Definition
EN 14511	COP	COP is defined as the ratio of the heat output of the heat pump unit to the effective energy input to the unit for a steady-state operating condition. Energy inputs and outputs are corrected for the pumping energy needed to overcome the pressure drop losses on the heat exchangers inside the unit.
	EER	Same definition as for the COP, used for cooling applications (useful energy is cooling).
EN 15879-1	COP/ EER	Uses same definitions as EN 14511, applied on direct expansion heat pumps.
EN 16147	COP	COP is defined as the ratio of the useful heat delivered for the production of domestic hot water and consumed electricity over a tapping cycle. The system includes the heat pump, the storage tank, and the circulation pump. Tank losses are accounted for.
AHRI 320/325/ 330	COP	COP is defined as the ratio of the heating capacity, excluding supplementary resistance heat, to the power input for steady-state operating conditions.
	EER	Same definition as for the COP, used for cooling applications (useful energy is cooling).

Table 4.A.1 (*Continued*)

Standard/guideline	PF	Definition
ISO 13256-1/ISO 13256-2	COP	COP is defined as the ratio of the net heating capacity to the effective power input of the equipment at steady-state operating conditions. The power inputs and outputs are corrected in the same way as in EN 14511.
	EER	Same definition as for the COP, used for cooling applications (useful energy is cooling).
EN 14825	SCOP	SCOP is defined as the ratio of the overall heating energy delivered over a 1-year time period to the total energy input to the system. It is a calculatory value obtained under certain assumptions regarding the heating load, climate, controls, and so on. The basis for the calculation is unit tests, for example, according to EN 14511.
	SEER	Same definition as for the SCOP, but for cooling applications.
ASHRAE 116	HSPF	HSPF is defined as the ratio of the total heat delivered over the heating season (not exceeding 12 months) to the total energy input over the heating season. It is a calculatory value obtained under certain assumptions regarding the heating load, climate, controls, and so on. The basis for the calculation is unit tests.
	SEER	SEER is defined as the ratio of the total heat removed during the normal period of usage for cooling (not exceeding 12 months) to the total energy input during the same period. Obtained similarly to the HSPF.
VDI 4650-1	SPF (e)	SPF(e) is defined as the ratio of the useful heat delivered in the period of 1 year over the electrical energy used to drive the compressor and some auxiliary drives. It is a calculatory figure based on the test results from EN 14511. It does not take into account electricity consumption for, for example, groundwater pump, heat pump off-mode, and so on.
EN 15316-4-2	SPF	SPF is defined as the ratio of the overall energy output to the overall energy input (final energy) of the heat pump system for heating and DHW. The heat pump system includes the heat pump unit, heat source, water storages, and all auxiliary systems (controls, liquid pumps, etc.).

4.A.2 Solar thermal collectors (Table 4.A.2)

CEN (2006) EN 12975-1:2006 – Thermal solar systems and components. Solar collectors. Test methods. CEN, Brussels, Belgium.

ISO (1995) ISO 9806-3 – Test methods for solar collectors. Part 3. Thermal performance of unglazed liquid heating collectors (sensible heat transfer only) including pressure drop. International Organization for Standardization, Geneva, Switzerland.

ASHRAE (2010) ASHRAE 93-2010 – Methods of Testing to Determine the Thermal Performance of Solar Collectors. ASHRAE, Atlanta, USA.

CEN (2012) EN 12976-2:2012 – Thermal solar systems and components. Factory made systems. Part 2. Test methods. CEN, Brussels, Belgium.

CEN (2011–2012) EN 12977 – Thermal solar systems and components. Custom built systems, Parts 1–5. CEN, Brussels, Belgium.

ISO (1993–2013) ISO 9459 – Solar heating. Domestic water heating systems, Parts 1, 2, 4 and 5. International Organization for Standardization, Geneva, Switzerland.

CEN (2007) EN 16316-4-3 – Heating systems in buildings. Method for calculation of system energy requirements and system efficiencies. Heat generation systems, thermal solar systems. CEN, Brussels, Belgium.

ISO (1999) ISO 9488:1999 – Solar energy. Vocabulary. International Organization for Standardization, Geneva, Switzerland.

Table 4.A.2 Overview of the performance factors for solar thermal collectors and solar thermal systems defined in different standards

		Solar thermal standards
Standard	PF	Definition
EN 12975-2	η	Collector thermal efficiency is the ratio of the energy removed by the heat transfer fluid over a specified time period to the product of a defined collector area (gross, absorber, or aperture) and the solar irradiation incident on the collector for the same period, under steady- or non-steady-state conditions (according to ISO 9488).
ISO 9806	η	Same as EN 12975.
ASHRAE 93	η_g	Collector thermal efficiency is defined as the ratio of the actual collected useful energy to the solar energy intercepted by the collector gross area.
EN 12976, EN 12977	f_{sol}	Solar fraction is the energy supplied by the solar part of a system divided by the total system load. The solar part of a system and any associated losses need to be specified, otherwise the solar fraction is not uniquely defined (according to ISO 9488).

Table 4.A.2 (Continued)

Standard	PF	Definition
	f_{sav}	Fractional energy savings is the reduction of purchased energy achieved by the use of a solar heating system, calculated as $1 - [$(auxiliary energy used by solar heating system)/(energy used by conventional heating system)$]$ in which both systems are assumed to use the same kind of conventional energy to supply the user with the same heat quantity giving the same thermal comfort over a specified time period (according to ISO 9488).
	Thermal performance	Thermal performance is defined as a set of performance indicators. For solar systems without auxiliary energy sources, these are the heat delivered by the solar heating system, Q_L; the solar fraction, f_{sol}; and the parasitic energy, Q_{par}. For systems including auxiliary energy sources, these are the net auxiliary energy demand, $Q_{aux,net}$; the fractional energy savings, f_{sav}; and the parasitic energy, Q_{par}.
ISO 9459	Thermal performance	Comparable definition to EN 12976 and EN 12977.
EN 15316-4-3		Same nomenclature as in EN 12977.

4.A.3 Relevant documents for the ecodesign directive

Commission Delegated Regulation (EU) No. 811/2013 of 18 February 2013 supplementing Directive 2010/30/EU of the European Parliament and of the Council with regard to the energy labelling of space heaters, combination heaters, packages of space heater, temperature control and solar device and packages of combination heater, temperature control and solar device.

Commission Delegated Regulation (EU) No. 812/2013 of 18 February 2013 supplementing Directive 2010/30/EU of the European Parliament and of the Council with regard to the energy labelling of water heaters, hot water storage tanks and packages of water heater and solar device.

Commission Regulation (EU) No. 813/2013 of 2 August 2013 implementing Directive 2009/125/EC of the European Parliament and of the Council with regard to ecodesign requirements for space heaters and combination heaters.

Commission Regulation (EU) No. 814/2013 of 2 August 2013 implementing Directive 2009/125/EC of the European Parliament and of the Council with regard to ecodesign requirements for water heaters and hot water storage tanks.

EN 14825:2012 – Air conditioners, liquid chilling packages and heat pumps, with electrically driven compressors, for space heating and cooling. Testing and rating at part load conditions and calculation of seasonal performance.

References

1. EU (2009) Directive 2009/125/EC of the European Parliament and of the Council of 21 October 2009 establishing a framework for the setting of ecodesign requirements for energy-related products (recast). *Official Journal of the European Union*, **L 285**, 10–35.

2. Wemhöner, C. and Afjei, T. (2003) Seasonal performance calculation for residential heat pumps with combined space heating and hot water production (FHBB method). Final project report within the research program Heat Pump Technologies, Cogeneration, Refrigeration of the Swiss Federal Office of Energy, Institute of Energy, University of Applied Sciences, Basel, Muttenz, Switzerland.

3. VDI (2004) VDI 6002, Blatt 1: Solare Trinkwassererwärmung – Allgemeine Grundlagen, Systemtechnik und Anwendung im Wohnungsbau, VDI, Düsseldorf, Germany.

4. Kramer, W., Oliva, A., Stryi-Hipp, G., Kobelt, S., Bestenlehner, D., Drück, H., Bühl, J., and Dasch, G. (2013) Solar-active-houses – analysis of the building concept based on detailed measurements. Proceedings of the International Conference on Solar Heating and Cooling for Buildings and Industry, September 15–23, Freiburg, Germany.

5. Heimrath, R. and Haller, M.Y. (2007) Project Report A2 of Subtask A: The Reference Heating System, the Template Solar System, Institut für Wärmetechnik, Graz University of Technology, Austria.

6. Frank, E., Haller, M.Y., Herkel, S., and Ruschenburg, J. (2010) Systematic classification of combined solar thermal and heat pump systems. EuroSun Conference, Graz, Austria.

7. OECD/IEA (2004) Energy Statistics Manual, OECD/IEA, Paris, France.

8. United Nations (2011) International Recommendations for Energy Statistics (IRES), Draft Version, United Nations, New York, USA. Available at http://unstats.un.org/unsd/statcom/doc11/BG-IRES.pdf.

9. EU (2006) Regulation (EC) No 842/2006 of the European Parliament and of the Council of 17 May 2006 on certain fluorinated greenhouse gases. *Official Journal of the EU*, **L 161**, 3.

10. Itten, R., Frischknecht, R., and Stücki, M. (2012) Life Cycle Inventories of Electricity Mixes and Grid, ESU-service (PSI), July 2012.

11. Ecoinvent (2013) International Database for Life Cycle Inventory Data, Swiss Center for Life Cycle Inventories, Dübendorf. Available at http://www.ecoinvent.org/database/.

5 Laboratory test procedures for solar and heat pump systems

Christian Schmidt, Ivan Malenković, Korbinian Kramer, Michel Y. Haller, Robert Haberl,
Anja Loose, Sebastian Bonk, Harald Drück, Jorge Facão, and Maria João Carvalho

Summary

Apart from the characterization of single components, laboratory tests on systems aim at obtaining the system performance figures "SHP" and "SHP+" that were introduced in Chapter 4. Several research institutes within T44A38 have been working on further development of test methods for their application on SHP systems. This work is summarized in this chapter.

The test procedures have been developed based on test procedures to a large extent originating from the field of solar thermal applications. They can be assigned to two main approaches (Section 5.2) that are named after the "test boundary" (Section 5.2.1): either the main performance-relevant components are installed in separate test rigs and measured individually (component testing and system simulation – CTSS) or all parts required for normal system operation are tested together (whole system testing – WST). Each approach has its own advantages and disadvantages, which are compared in Section 5.2.2. Section 5.2.3 summarizes the findings of an investigation about the scope of these newly developed test procedures. Although there are limitations, already more than 70% of the market-available solar and heat pump (SHP) system configurations can be characterized with these methods today. Measuring components or systems for a certain period at a test rig is one thing, but obtaining the annual performance figures is another important task. Most test procedures rely on modeling and simulation, which is bound to certain drawbacks if appropriate models are not available. However, for whole system test methods also a direct extrapolation method is available (Section 5.2.4). Apart from performance figures, further possible outputs of testing (Section 5.2.5) are quality and performance labels. For product developers, testing may also return valuable information on malfunctions, design faults, or weaknesses, and lead to subsequently improved products. In Section 5.3, experiences from testing of SHP systems with the new or further developed test methods are reported. Section 5.4 contains a summary and conclusions.

In this chapter, "test procedure" refers to a concrete instruction on how a test of an SHP system is to be conducted in a laboratory, while (test) "method" refers to the approach and the general idea behind test procedures.

Solar and Heat Pump Systems for Residential Buildings, First Edition.
Edited by Jean-Christophe Hadorn.
© 2015 Ernst & Sohn GmbH & Co. KG. Published 2015 by Ernst & Sohn GmbH & Co. KG.

5.1 Introduction

Sustainable growth and market penetration often rely on standardized, affordable, and transparent measures of quality assurance. If disregarded, consequences might be tremendous. A prominent example is the heat pump market development and its abrupt breakdown in the 1980s due to lack of quality assurance combined with poorly conceived system concepts with design faults and faulty installations. Since then, heat pump technology has become a mature technology with increasing market shares in markets around the world. In this context, quality assurance tests provided by accredited test institutions played a pivotal role. Thus, products are tested by an institution whose credibility is certified and controlled by an independent accreditation body. Crucial for transparent quality assurance methods is the availability of standardized testing and rating procedures. These procedures have to objectively promote high-quality and efficient products without favoring different technical solutions.

These test procedures constitute the basis for energy efficiency standards and labels such as the EU Ecodesign and Energy Labeling Directives[1)] and, ultimately, also for politically motivated incentive programs such as tax deductions or subsidies. In the case of room air conditioners, history has shown that, if well implemented, such measures can effectively support a market transformation toward more energy-efficient products [1].

In this context, the question arises what actually constitutes an effective test procedure that can be accepted by all relevant stakeholders as a standard. An example for an internationally well-spread whole system test procedure is the NEDC (New European Driving Cycle) in the automotive sector, which originated from a German DIN standard. It is nowadays accepted by many countries worldwide. This test cycle is carried out on a roller test bench to assess the fuel economy and the level of emissions of cars.

It can be generally stated that test methods and procedures have to meet different and partly contradicting requirements posed by stakeholders such as manufacturers (industry), test institutions, end customers, and policy makers (cf. Table 5.1). For example, looking at the NEDC, the method is very flexible in its application to different types of cars while high repeat accuracy and repeatability can be achieved by the automatic control of the roller test bench. Since the test boundary conditions are well controlled, comparability of the determined performance figure is high among different products (cars) and different test laboratories. On the other hand, NEDC is criticized since the cycle is far too simple to represent real driving (poor informative value). Here lies the strength of real-world driving cycles, which, however, have weaknesses in the aforementioned criteria. The mentioned principles and the contradicting requirements apply also for testing procedures for different energy using products and, in particular, to test procedures for SHP systems and components.

Ideally, a testing and rating standard for SHP systems should make it easy for potential buyers, planners, installers, and policy makers to compare relevant properties of SHP

[1)] The reader is invited to refer to Section 5.4 for its implications on SHP systems.

5.1 Introduction

Table 5.1 General requirements for test methods (partly adapted from Ref. [2]) and relevance for selected target groups

Requirements for test procedures		Relevance for industry (I), test institutes (T), customers (C), and policy makers (P)
High degree of	Description	
Comparability	The figure(s) of merit identified allow(s) comparison with similar types of products.	Especially valuable for (C), can provide a basis for subsidy schemes (P), and (I) can improve products (e.g., before–after tests).
Informative value	The test should reflect actual usage conditions and the result should be valid for a broad range of boundary conditions.	Test methods with high degree of comparability and informative value are able to improve the transparency in the market → all stakeholder groups can benefit.
Simplicity and clarity of definition	Test procedures should be clearly described, so that they can be implemented in the same way by all test bodies. Furthermore, precise definitions can prevent manufacturers to use technical loopholes to influence test results.	Well-developed test methods allow quick standardization and introduction of labeling scheme/manipulation-proof test methods → likewise relevant for all stakeholder groups.
Repeatability	Low degree of uncertainty of the result of the same product when multiple tests are carried out by different institutes.	A certain minimum of accuracy that has to be fulfilled → likewise relevant for all stakeholder groups.
Flexibility	Flexibility in terms of applicability not only to different system configuration types, but also to new technologies; extrapolation of the test results to similar system configurations ("system families").	(T) can offer a broad range of tests. Extrapolation of test results can significantly reduce cost of testing for (I). Likewise relevant if considering standards, since adjustments of standards are a drawn-out process.
Cost effectiveness	Refers to short test duration and low requirements of both skills of the operating staff and regarding test laboratory infrastructure.	(T) can benefit due to higher profit margins. Lower overall costs enable smaller companies (I) to buy tests.

systems among themselves, as well as with other technologies that serve the same purpose. A good example is the energy labeling for various household products that is already applied in many markets around the globe. Similarly to the NEDC example, it is not an easy task to define a versatile test method and to derive testing and rating standards for whole SHP systems. As stated in Chapter 4, there are currently no widely accepted *standardized* testing and evaluation methods for SHP systems.

Current challenges in developing test procedures for SHP systems are based on the fact that test procedures have been developed to characterize the single technologies (solar thermal systems and heat pumps) and there is no obvious solution on how to integrate the one or the other approach into a common procedure. The concrete deficiencies of these standards and guidelines for the application on SHP systems have been identified within the European QAiST project [3]. In conclusion, it can be said that current solar thermal standards cannot characterize the heat pump as an additional heat source. On the other hand, current heat pump test standards are component tests that do not apply to systems.

However, prior to and especially during T44A38, efforts have been made to further develop solar thermal system test procedures in order to include heat pumps in these tests. Due to the fact that the solar thermal system is usually combined with other means of heating, solar thermal test procedures have been dealing with additional heating elements within their test procedures already for a long time.

However, compared with the single technologies, system complexity is significantly higher due to the variety of possible system concepts (compare Chapter 2). Moreover, compared with, for example, gas boilers combined with solar thermal systems, different and sometimes several heat sources of a heat pump need to be considered in the design of both the test rig and the test method. Finally, more sophisticated control strategies need to be considered.

There are currently several test procedures for solar thermal systems that can be reduced to two basic approaches: component testing and system simulation and whole system testing. Several research institutes within T44A38 have been working on further development of these approaches for their application on SHP systems. This work is summarized in this chapter. In Section 5.3, first experiences from laboratory testing of SHP systems are described.

5.2 Component testing and whole system testing

5.2.1 Testing boundary and implications on the test procedures

SHP test procedures can be subdivided into test procedures that can be applied to systems for domestic hot water (DHW) preparation and combi-systems that provide DHW and space heating. Some SHP systems also provide active or passive cooling. However, none of the test procedures discussed in this chapter has assessed this feature yet. According to the basic methodology, all described procedures can be divided into "component-based testing" and "whole system testing."

5.2 Component testing and whole system testing

Figure 5.1 shows the main methodological differences in terms of testing boundary and long-term performance evaluation approaches, both for combi-systems and for DHW-only systems.

In case of CTSS, each component is tested individually according to a separate methodology described in respective normative documents (refer to Section 5.2.2 for details regarding the test procedures). WST methods basically require installation of the whole system at the test rig, as provided by the manufacturer, and the tests are performed on the system as a whole.

This implies that the system boundaries of the equipment under test differ to a great extent for the two described methods. In case of WST, the system boundary for testing corresponds to the system boundary for the evaluation (see Figure 5.2: system boundary for testing). Apparently, this is not the case for CTSS, where the system performance figure (based on a boundary of the entire system) is evaluated out of results of individual component tests. Hence, interconnecting components are not included in the test but still have to be considered somehow in the system performance assessment.

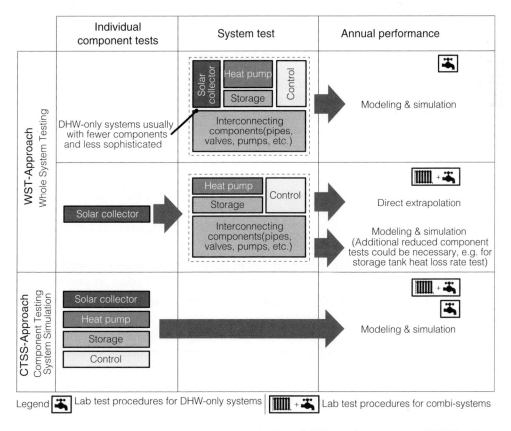

Fig. 5.1 Test methods available for the characterization of SHP combi-systems and DHW-only SHP systems

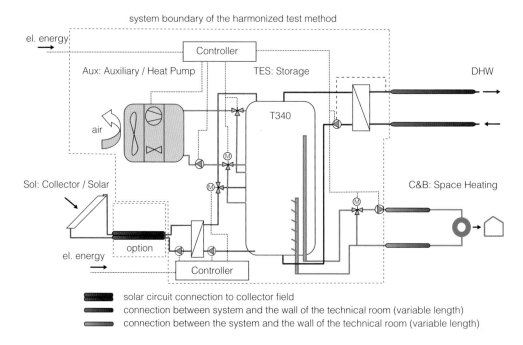

Fig. 5.2 Boundary of WST for combi-systems (harmonized test procedure according to Ref. [4])

A commonality for CTSS and WST is that the large heat pump ground heat exchangers cannot be installed in a laboratory for economic reasons. Hence, they are being emulated and simulated using standard boundary conditions. External units of air source heat pumps, however, can be installed and measured using a climate chamber (e.g., as indicated in Figure 5.2 for WST for combi-systems). As can be seen in Figure 5.1, WST procedures for combi-systems exclude the (large) solar collector area, while WST procedures for DHW-only systems include the (much smaller) solar collector area in the system test boundary. This means that for combi-system WST the collector must have been previously characterized. Simulation and emulation of the collector field, using feedback from the equipment under test in a hardware-in-the-loop procedure, allows for proper considering of the solar collector field behavior.

Usually, measurements inside the boundary of the tested system are not carried out. In case such measurements are still necessary, for example, in order to provide more detailed feedback to the manufacturer, possible effects introduced by sensors must be either negligible or compensated for.

While WST may include the space heat distribution pump and supply temperature mixer, neither CTSS nor WST includes other parts of the distribution system for space heating and DHW within the testing boundary. The reason is that these usually differ in every house, and they are usually not part of the package sold by the system manufacturer. However, it should be noted that the actual distribution system design and use pattern can have tremendous influence on the system performance. The heat losses of the supply and return pipes from the collectors are included in some of the procedures for WST. For CTSS, these may only be included in the simulation of the whole system.

Looking at the different testing boundaries of CTSS and WST, it has to be kept in mind that ensuring the same system boundary for the performance evaluation is an important requirement to provide comparable testing results for CTSS and WST. Hence, for CTSS, the energetic influence on the system performance of the inter-connecting components such as pipes (i.e., heat losses), pumps (auxiliary energy consumption), and valves (energy flows) – which do not form part of the testing boundary – still need to be considered in the performance evaluation. Correct and appropriate consideration of such energetic influences for all kinds of system configurations without consideration in a test is principally difficult. In CTSS, heat losses of pipes and auxiliary energy consumption of pumps can be calculated by means of simulation, if the controller has been previously tested and the controller strategy is appropriately represented in the system model (cf. EN 12977-2 [5]). On the other hand, simplifications usually have to be applied when mapping the controller or in some cases even the hydraulics in simulation. Also, sometimes it is not possible to get the required information about the control strategy from the manufacturer. In this case, a generic control strategy is assumed and the test needs to prove that the applied strategy from the manufacturer does not perform worse. However, the standard does not clarify how to prove this.

Regarding the assessment of the controller behavior, it should be considered that the controller in a real system is basically interacting with the building (and its users) and therefore with external effects related to its energy consumption, such as passive (solar) gains, internal energy gains, thermal inertia of the building and different heating zones, and the comfort level. By setting up an annual simulation with a fixed load profile (CTSS), these effects are not further regarded as they can be included in the profile. In case of WST for combi-systems, the building load is dynamically simulated; hence, such feedback from the building and the response of the controller can be evaluated during the test (hardware-in-the-loop). In this way, controller behavior can be checked in a very holistic manner. The drawback of the highly autonomous controller behavior is that the consumed energy, a value directly correlated to the final performance figure, will be determined by the controller as well. As a result, a fair comparison of several systems with direct extrapolation (cf. 5.2.4) becomes difficult, since controller settings may become the main determinant for the system's energy demand and hence its performance. Recent developments within the EU research project MacSheep (www.macsheep.spf.ch) showed that it is possible to simulate and emulate the response of the building, while at the same time guaranteeing identical amounts of heat delivered to the building even with different controllers and control settings.

5.2.2 Direct comparison of CTSS and WST

Table 5.2 gives an overview of the major differences, advantages, and disadvantages of the two methods: CTSS and WST. Listed statements for WST procedures are principally

Table 5.2 Comparison of methods to characterize the performance of SHP systems (DHW-only and combi-systems). There are limitations regarding the statements made for WST procedures for DHW-only systems.

	CTSS	WST
Why perform?	– If concise characterization of a component is of particular interest (e.g., assess variation of certain design parameters on component performance). – Thermal storage and collector test results can be used to obtain Solar Keymark and EU energy label. – Suitable for performance characterization of *customized systems* set up by manufacturers or retailers.	– If proper system function is of particular interest, for example, to check proper behavior of components and their interactions. – Assure correct system function, prior to/instead of field testing of prototypes. – Evaluation of the effect of component and control changes on *system performance*, that is for system *development/R&D*. – Suitable for highly *prefabricated systems* with high product sales.
Philosophy	– Cost reduction of testing larger product portfolios by allowing identical components employed in several systems to be tested only once (relevant for labeling schemes).	– Cost reduction and more reliable performance information for highly integrated systems that are always sold in one package (not custom built).
Advantages	– Allows manufacturers with a larger variety of system configurations to reduce testing costs. – If several products are to be tested, only differing or missing system components need to be tested (not all individual system configurations) because they can be evaluated by means of simulation. – Precise parameterization of tested components allows flexible performance evaluation for many load patterns and climates by simulation under given assumptions. Such simulation studies based on experimental data can help to optimize design parameters.	– Only by testing the system as a whole proper function of all components can be checked. This includes thermal storage stratification and heat losses, heat pump starts, stops, and defrosting, solar group, valve leakage, pipes, control, etc.) under various boundary conditions – Knowledge of the manufacturer's control strategy is not needed. – Effective storage heat trapping and prevention of unwanted gravity-induced circulation can be checked and reported. – In case of *direct extrapolation* (cf. Section 5.2.4) of the test result to annual performance

Table 5.2 (Continued)

	CTSS	WST
	– Certain scaling options available: Calculation procedures to transfer test results of collectors and storages to similar products of different size. – Thermal performance can be more flexibly evaluated for any arbitrary boundary conditions (climate data, heating, and DHW load).	figures, the method does not depend on the availability of simulation models for the tested components to obtain the final performance figure; hence, it is open to any new technology without additional efforts for component characterization and modeling. – More accurate *system* performance evaluation.
Disadvantages	– For every component to be tested, an appropriate model has to be available. This is often an issue if new innovative systems are to be characterized. – Shortcomings of models – for example, the incapability of currently used 1D storage models to simulate stratification accurately introduces additional uncertainties. – The result of the long-term performance evaluation simulation can fairly overestimate actual performance due to undetected malfunctions, which occur only when the system is operated as a whole.	– Extrapolation to other boundary conditions is yet a difficult task where research is ongoing. – Not (yet) integrated into quality and performance labeling schemes. – No scaling available yet.
Related test procedures	– CTSS is based on a set of mostly standardized component test procedures applicable to various types of "DHW-only" and combi-SHP systems: Collector: ISO 9806 [6]; storage tank: EN 12977-3,4 [7,8]; control: EN 12977-5 [9]; for the heat pump, a standard does not exist (yet). A procedure is under development (see Section 5.3.1).	– For various types of SHP combi-systems: "Harmonized WST" is under development [10]. – For various types of DHW-only SHP systems: Further developed test procedures for solar water heaters [12,13] are based on the "Dynamic System Test" (ISO 9459-5 [11]). [14] (cf. Section 5.3.3) is a further development of the former heat pump water heater test EN 16147.

valid for all related procedures (DHW-only and combi-systems) apart from the following limitations applying for test procedures for DHW-only SHP systems:

- The direct extrapolation approach is not available (cf. 5.2.4).
- Instead of "close to reality" test sequences, mostly "parameter fitting" test sequences are applied (cf. Figure 5.3), which – due to the different test sequence characteristics – make it more difficult to spot system malfunctions during testing, but on the other hand allow better extrapolation possibilities.

The CTSS method has been recently standardized. So far, it applies to solar thermal systems only; SHP systems have not been covered yet. This is due to the fact that the heat pump component test procedure was still under development at the time this handbook was written (see also Section 5.3.1). Concerning the whole system test procedures for combi-systems, several methods were developed in the past [15–17], but none of them has been standardized so far. Within the EU project MacSheep, the current state of the art of WST procedures for combi-systems has been summarized and discussed [10], and a harmonization of the different methods has been started.

5.2.3 Applicability to SHP systems

Within T44A38, an investigation was performed in order to systematically assess the applicability of the available test procedures to the marketed SHP systems and identify their strengths and weaknesses.

The assessment considers the SHP systems' main features, including the system concept (parallel, series, and regenerative) and the possible energy flows as depicted in the energy flow chart (see Section 2.1). Further focus was on the applicability to systems containing "extraordinary" components – special types of collectors (e.g., PVT), certain storage technologies (e.g., latent heat storages), and new heat pump technologies (e.g., desuperheater). Basis for the investigation was T44A38 market survey presented in Section 2.2.

In Table 5.3, the main results of the assessment are summarized. Looking at the first row, it can be seen that further differences in terms of applicability do not depend on the method, but on the test data evaluation methods applied for the long-term performance prediction (cf. Section 5.2.4). Two methods are found: modeling and simulation (applied in CTSS and WST) and direct extrapolation (only applied in WST, cf. Figure 5.1). It can be concluded that both CTSS and WST methods applying modeling and simulation for the performance evaluation have some restrictions regarding applicability to SHP systems containing certain types of components because of lack of appropriate simulation models. For direct extrapolation of the test results to obtain seasonal or yearly performance, there are no restrictions on system configuration or applied components, which can be a great advantage in a fast-changing energy supply environment.

Limitations of CTSS and WST are listed in Table 5.4. The table should serve as a first orientation. Since not every system could be investigated in every detail, further limitations could arise that might impede a full characterization of a given system as intended by the test procedure. Concerning the applicability of the test data evaluation methods for DHW-only SHP systems, for CTSS, the same scope and the same limitations

Table 5.3 Applicability of the two test data evaluation methods on different SHP combi-system configurations

Test data evaluation method	CTSS and WST: modeling and simulation	WST: direct extrapolation
	Scope of test methods for solar and heat pump systems	
Concept	Serial, parallel, and combined serial and parallel.	
Heat sources	Solar, ambient air, ground, water, nonventilated exhaust air (e.g., basement/boiler room).	
Heat sinks	DHW and space heating with water-based heat distribution.	
Collector	All types of flat-plate and vacuum collectors that can be fully characterized according to EN ISO 9806 [6] and that can be represented in simulation; furthermore, unglazed and PVT collectors able to use ambient air.	
Heat pump	Fixed capacity and capacity-controlled (with variable-speed compressors) heat pumps, regardless of the refrigerant, with and without electric backup heater. Also systems employing heat pumps with low-lift compressors can be characterized.	Since no model is required, the method is applicable on all types of heat pumps including, for example, heat pumps with economizer, subcooler, or desuperheater.
Storage	Typical types such as combistore, DHW, and buffer to be installed in the basement/boiler room (e.g., below 3 m^3 according to EN 12977-3,4) that use water as storage medium, but also brine as a storage medium is principally possible. Most stratification/charging/discharging devices can be modeled – with some limitations in modeling accuracy.	Systems using thermal storage tanks with innovations that cannot be modeled accurately yet can be evaluated: For example, highly integrated systems where components cannot be separated (e.g., storage with integrated boiler or heat pump) or integrated thermal PCM storages.
Control	Testing of controllers and associated equipment includes sensors (in particular to measure temperature, irradiance, pressure, flow, or heat), actuators (pumps, solenoid, or motor valves and relays). Refer to Table 5.4 for methodological limitations and differences of the two test data evaluation methods on the scope.	

Table 5.4 Known limitations of the two test data evaluation methods on different SHP combi-system configurations

Test data evaluation method	CTSS and WST – modeling and simulation	WST – direct extrapolation
Limitations of test data evaluation methods for solar and heat pump systems		
Concept	Regenerative (any SHP concept with ground temperature regeneration) can only be assessed by means of simulation/modeling of the ground source.	
Heat sources	Exhaust air collected within a ventilation system. This could be integrated into all systems, but has not been done so far. Direct evaporation earth collectors.	
Heat sinks	Cooling (active and/or passive), systems with air as a means of energy distribution, systems comprising an additional tile stove/open fireplace	
Collector	Collectors that cannot be fully characterized according to ISO 9806, such as hybrid collectors (combined liquid/air) and other "extraordinary" collector types. For SHP systems with PVT collectors, the electric performance cannot be fully assessed.	
Heat pump	To date of publication, heat pumps with economizer or desuperheater cycles have not yet been modeled and parameterized with test data (within the frame of solar and heat pump system testing).	No limitations known.
Storage	Large and/or buried thermal storages that cannot be installed in a laboratory. Seasonal storage concepts that comprise, for example, chemical storage materials.	
	Systems with storages, for which appropriate models are not yet available: See Table 5.3, column "WST – direct extrapolation", row "storage" for examples.	No further limitations known.
Control	System controllers comprising forecast strategies (weather forecast control or user profile learning algorithms) cannot be characterized. Also, if the heating system is operated by a superior building control (e.g., home energy management control in combination with a PV array), it cannot be characterized.	

Table 5.4 (*Continued*)

Test data evaluation method	CTSS and WST – modeling and simulation	WST – direct extrapolation
	Limitations of test data evaluation methods for solar and heat pump systems	
	Representation of the exact controller strategy and behavior within a system model is usually not possible due to the effort involved. Furthermore, manufacturers might not be willing to disclose precise controller operation algorithms. Simplified strategies can be used to obtain long-term performance while the test needs to provide evidence that the tested SHP will not perform inferior in reality (however, EN 12977-5 lacks information on how to perform this step).	No further limitation. Control is assessed "as is" and does not rely on modeling or disclosing control algorithms by the manufacturer.

apply as for combi-systems. Looking at the WST test procedures for DHW-only systems, the available procedures are comparably new. Two test procedures can only be used with serial DHW-only system concepts where the collector directly acts as the evaporator of the heat pump ([12,14], cf. Section 5.3.3). The test procedure presented in Ref. [13] can be applied on systems where the condenser of the heat pump is either immersed into the tank or externally placed. Theoretically, also the old test procedure described in ISO 9459-5 could be applied on SHP systems. However, the test procedure is unlikely to reveal realistic results, since the heat pump is not installed for the test.

In conclusion, it can be stated that more than 70% of the system configurations identified in the market survey (cf. Section 2.2) could be tested with at least one of the SHP test procedures given in Table 5.2. This considers the status of these test procedures (cf. Table 5.2, "Related test procedures") at the end of 2013.

5.2.4 Test sequences and determination of annual performance

Determination of long-term performance of SHP systems is based on measurement data postprocessing. Resulting performance figures are – apart from the characterization of single components – "SHP" and "SHP+" (cf. Chapter 4). As mentioned in Sections 5.2.1 and 5.2.3, two test data evaluation methods for the determination of annual performance can be found in test procedures for SHP systems: "Modeling and

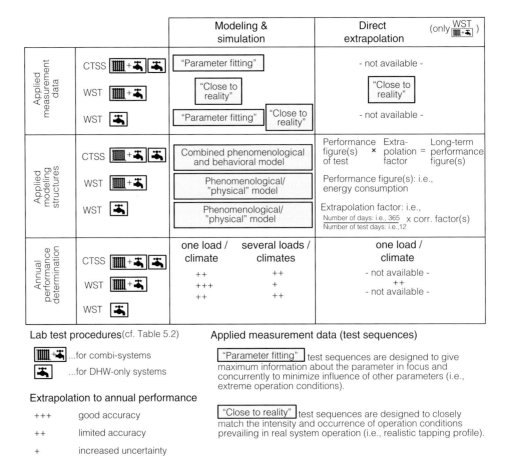

Fig. 5.3 Implications of type of test sequences and applied modeling structure on the long-term performance prediction

simulation" and "Direct extrapolation". Figure 5.3 summarizes the applied types of measurement data (directly related to the types of test sequences), modeling structures, and the flexibility of the annual performance determination applied in both test data evaluation methods.

5.2.4.1 Direct extrapolation of results (WST for combi-systems)

Direct extrapolation was originally developed in Ref. [17] for solar thermal combi-systems. It is based on a single "close to reality" test sequence, which is composed of a number of consecutive test days, each representing a "real case day," for example, two winter days, two summer days, and two days of the transitional period, whereas each day has a realistic tapping profile, space heating load, and typical ambient air temperatures/global irradiation conditions. Hence, this continuous test sequence is designed to match closely the intensity and occurrence of operating conditions prevailing in real system operation.

5.2 Component testing and whole system testing

During the test, the performance of the tested system is measured (i.e., auxiliary energy consumption). In order to obtain annual values of the performance figures, formulas are applied (see example in Figure 5.3).

Direct extrapolation is applied in the "harmonized WST procedure" [10] applicable to SHP combi-systems. The major advantage of this type of long-term performance evaluation is that no simulation model is required to obtain long-term values, which is a major difference to all other test procedures discussed here. The disadvantage is that a flexible extrapolation to other boundary conditions (load profiles and climates) is difficult, because the results for these boundary conditions may depend on operating conditions that were not included or not adequately represented in the test sequence that was applied (cf. Figure 5.3, direct extrapolation: annual performance determination only validated for one climate and load). Further work will be needed in order to find simple extrapolation methods that do not require detailed models. On the other hand, also the driving cycle tests from the automotive sector are lacking extrapolation to other boundary conditions and have been accepted widely.

5.2.4.2 Modeling and simulation

Two types of model structures can be seen in current SHP test procedures: "phenomenological" (or physical/knowledge-based/white box) models and behavioral (sometimes "black box") models. Test sequences used to identify parameters of phenomenological models (Table 5.5) usually consist of several test sequences, each designed to identify one model parameter while minimizing the influence of all other parameters. This can lead to test sequences, which do not necessarily have to occur during real system operation, that is, with extreme operating conditions. Principally, also "close to reality" test sequences (cf. Figure 5.3) can be suitable to determine model parameters; however, determination of parameters (parameter identification) can be more challenging, since parameter correlations can occur more easily.

Test sequences as well as the availability of a model that is principally able to represent the energy flows of the real system are the basis for the parameter identification procedure. Thus, all dimensional parameters of the system (WST)/component (CTSS) under test are set up in the model. A number of performance-relevant model parameters are then automatically determined using simulation software coupled to an optimizer. The optimizer runs numerous simulations where the performance-relevant model parameters are varied, while the summarized difference of all simulated and measured output values – usually transferred energy/power – is minimized. Parameter identification is finished, when this difference is minimal (and below a specified threshold). In this case, the model is able to reproduce the transferred power and energy of the equipment under test during the measurement sequence. Depending on the method, the model can also represent system performance under enhanced boundary conditions to a different extent[2] (Figure 5.3). Evaluation of system performance under various boundary

---

[2] The question to which extent system performance under *enhanced* boundary conditions (several loads/climates) can be evaluated with a given test procedure is also subject to the validation activities (monitoring of a real system) during its development.

Table 5.5 General differences between phenomenological and behavioral models

	Phenomenological model	Behavioral model
Basis	Mathematical equations based on physical/observed relationships	Mathematical equations based on function of (biological) neural networks (brain)
Examples in SHP systems	Solar collector equation (e.g., as in EN ISO 9806), simulation model of a thermal storage tank (e.g., TRNSYS Type 340 [19])	Current heat pump model for CTSS as described in Section 5.3.1
Parameters	Concrete meaning, that is, heat loss rate parameters of a collector or storage	No concrete meaning related to the object in focus, for example, "synaptic weights"
Prior information	Taken into account, that is, physical relationships in form of mathematical equations	Neglected, model based only on measurement/training data
Validity domain when extrapolation to other boundary conditions	If model is correct, large validity domain beyond conditions prevailing during testing	Since no prior information given, validity beyond testing conditions usually very poor
Simulation effort	Usually requires significantly increased computational power compared with behavioral models	Model training potentially time consuming, but simulations quick due to simple model structure

Adapted from Ref. [18].

conditions allows, for example, optimizing system design parameters. The favorable flexibility comes along with the disadvantage of increased effort in detailed component characterization compared with direct extrapolation methods (in terms of postprocessing of test sequence results). In this context, proper implementation of the controller functions into the system model can be a critical issue (see also Section 5.2.1, last para.).

While the phenomenological models require correct implementation of all relevant physical processes (e.g., heat transfer, pressure drop, and two-phase flows), behavioral models for complex systems require measurements (training data) under a much larger number of "system states." Especially thermal storage and controller, as highly dynamic components, drive the number of system states to be tested to an economically unfeasible extent; hence, pure behavioral system models are not suitable. Instead, phenomenological and behavioral models can be combined ("gray box" models) in order to obtain advantages of both model structures while avoiding their drawbacks. Thus, simple phenomenological equations that represent the energetic behavior are used where models are already available, while behavioral structures are applied where models are missing or where modeling becomes too complex. Such a combined system model has been reported in Ref. [17] for solar thermal systems. While CTSS hitherto relied on purely phenomenological models, within T44A38 a behavioral model was developed also for the heat pump (cf. Section 5.3.1).

5.2.5 Output

Results of performance test procedures may be indicators for the overall system performance or other test results that give useful information either for performance rating or for system development (Figure 5.4).

For the salesman, only performance indicators that can be used for marketing are of interest. These are usually few values (the fewer the better) that can be explained easily to the costumer. However, for the system developer, other indicators may be of interest for further R&D.

A key driver for manufacturers to request testing is usually to obtain widely accepted and trusted labels for their products. Such quality and/or performance labels can be a "must have" in order to enter certain markets, mostly for regulatory reasons.

Another output of performance testing is information about quality issues and operation abnormalities/malfunctions that influence the system performance. Such information obtained during installation, start-up, and testing is especially valuable for manufacturers, since they can improve their products. In this context, the relevance of labels that not

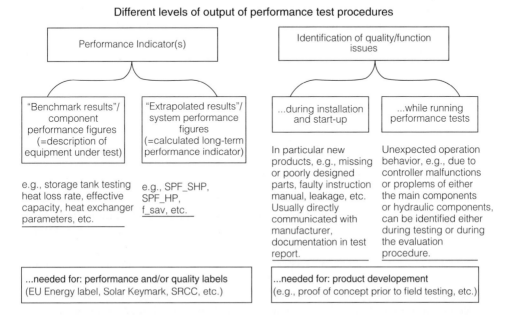

Fig. 5.4 Different levels of output of performance test procedures

only account for the energetic performance (i.e., EU energy label) but also define certain quality requirements (e.g., Solar Keymark) shall be mentioned.

CTSS and WST are principally both suitable to identify performance figures and quality issues (keeping in mind the advantages and disadvantage from Table 5.2). Operational faults are more difficult to detect when applying CTSS, because the control is tested individually without being able to receive feedback from affected components. For example, the position of certain connecting ports and sensors as well as the stratification efficiency of the storage may lead to significant performance loss for SHP systems [20,21].

5.3   Experience from laboratory testing

5.3.1   Extension of CTSS test procedure toward solar and heat pump systems

The CTSS method is a laboratory test method originally developed for the determination of the annual thermal performance of solar thermal systems. Since then, it has been applied to a large number of solar thermal domestic hot water systems and combi-systems with fossil or electric backup heater.

In 2007, first solar and heat pump systems were investigated based on the CTSS method [22], where the heat pump was modeled based on a performance map for steady-state operation with a fixed temperature difference between the inlet and outlet temperatures of the heat pump. However, since the operating conditions of combined solar and heat pump systems are often significantly different from steady-state behavior and of rather transient nature, this assumption leads to inaccurate long-term performance evaluation results. Hence, a dynamic test method with transient operating conditions for the heat pump as component of SHP systems was developed within the German national project WPSol [23] and experimentally applied to a brine to water heat pump. Based on the hereby acquired laboratory test data, an artificial neural network (ANN) was trained in order to characterize the thermal performance of the heat pump under dynamic operating conditions.

As shown exemplarily in Figure 5.5, during the two operation cycles the calculated outlet temperatures (T_prim/sec,out_sim) as well as the calculated electric power (P_el_sim) correspond very well with the measured temperatures (T_prim/sec,out_-meas) and electric power (P_el_meas) of the heat pump.

With the trained ANN for the heat pump and numerical models for all other core components of an SHP system that are tested conventionally according to EN 12977, the annual thermal performance of the overall system can be predicted for defined reference conditions (meteorological data and load profiles) by using a component-based simulation program such as TRNSYS. For the validation of the new test method, the results have been compared with measurement data from a field test in which the same type of heat pump has been used. In a second laboratory test, an air to water heat pump has been examined at TZS/ITW in a similar way as the brine to water heat pump, using a climate chamber for defined air temperature and humidity.

5.3 Experience from laboratory testing

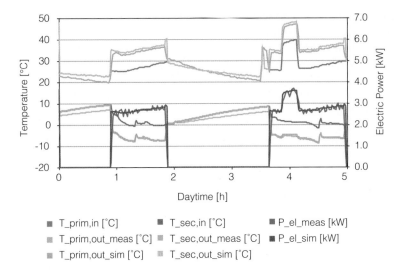

Fig. 5.5 Extract of the validation data showing the measured inlet and outlet temperatures on the primary (heat sink) and the secondary side (heat source) of the heat pump (left axis) and the electrical power of the heat pump (right axis) in comparison with the corresponding simulated quantities during two operation cycles

The clear advantage of the new heat pump test procedure implementing artificial neural networks is its potential for the applicability to a very wide range of heat pump products, because it is not based upon a physical model. Thus, it is not bound to the limitations of a specific set of equations. On the other hand, the usage of the ANN model completely lacks the ability for extrapolation, which is the most significant disadvantage. If the operating conditions during simulation extend the boundaries of the data used for training of the ANN, the calculated results become meaningless. Therefore, more work will be undertaken in order to extend the ANN method to so-called model-based artificial neural networks, that is, a combination of physical models with the ANNs, which are more convenient for a broader application.

Since it has been shown that the developed test method is very suitable for the characterization of the thermal performance of the heat pump integrated into an SHP system, the next step is to include it into the standard series EN 12977. The EN 12977 series contains test procedures for the components of a solar thermal system (e.g., heat store and controller) in a separate part of the standard, each. Therefore, it is planned to introduce the "test procedure for electrically driven heat pumps for thermal characterization with an artificial neural network" as an additional part (e.g., Part 6) of the standard. To do so, activities will be initiated in the European CEN TC 312 WG 3 after the completion of the project WPSol.

5.3.2 Results of whole system testing of solar and heat pump systems

The concise cycle test (CCT) method is a whole system test method that was developed for the evaluation of the performance of complete heating systems in a laboratory under realistic operating conditions ("close to reality" test sequences, cf. Figure 5.3) [15]. The method has been successfully applied for systems that combine solar thermal collectors with oil, gas, and pellet boilers [24,25]. Within the Swiss national project SOL-HEAP, the method has been extended to systems with heat pumps [26]. A total of 11 systems that combine solar thermal and heat pumps have been tested. Common features of the tested systems are that they combine solar thermal collectors, a combistore, and a heat pump. Whereas seven of the tested systems were pure parallel system concepts where the solar thermal collectors deliver heat to the combistore only, four system tests have been performed with concepts where heat from the solar collectors is used not only for charging the combistore, but also for the evaporator of the heat pump. Two system tests were performed with systems where heat from the solar collectors was the only heat source for the heat pump.

The 12-day test procedure proved to be applicable to all system designs. The achieved performance results were sometimes quite worse than the manufacturers expected. Thus, the second step after the laboratory testing, the determination of annual performance figures with the help of annual simulations, was only performed for some of the systems. More information about the test procedure and results can be found in Refs [20,26].

The performance factors achieved during the 12-day tests range from 2.7 to 4.8. However, a direct comparison between the test results is of limited use because the space heat loads were not identical for all cases. The reason for this is that some of the systems have been designed exclusively for passive houses and were tested with a building emulation that corresponded to a passive house, whereas other systems were tested with buildings that had a design annual heat demand of 60–100 kWh/m^2. A second reason was that even for identical building definitions the space heat delivered differed substantially because the tested systems were allowed to deliver heat according to their own controllers.

This is illustrated by Figure 5.6 showing performance factors of a system that was tested twice with different solutions for hydraulics and control. The performance factor was thereby improved from 4.0 in test A to 4.8 in test B. However, despite the better performance factor, the electricity consumption of test B was higher than that for test A. The higher performance factor is a result of more energy delivered for space heating due

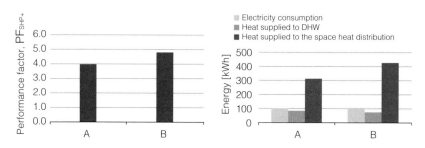

Fig. 5.6 Performance factors of the 12-day test of a system in two versions (A and B). Version A has a higher PF_{SHP+} and at the same time higher electricity consumption than version B

to wrong controller settings that enabled the system to deliver heat to the building in summer. This heat was generated under very favorable conditions, but it was useless because there was effectively no heat demand.

In the following section, two problems that have been found in some of the tested systems are illustrated with results from measurements.

5.3.2.1 Excessive charging of the DHW zone

Figure 5.7 shows exemplarily the progress of the 5th test day of a parallel system with flat-plate collectors and a ground source heat pump [27]. The HP is running in space heating mode at the beginning of the day and charging the space heating zone of the storage in an oscillating on/off mode, which can be seen from the oscillation of the temperature sensor T_{S5}. Around hours 15–18, the HP is charging the upper part of the store that is reserved for DHW preparation. It can be seen that the temperature in a region of the tank that was used for space heating before (TS5) rises to 50 °C, and even the temperature at the position T_{S4} rises to 40 °C. For the rest of the day, the heat pump does not have to deliver low-temperature space heat anymore since heat is taken from the excessively heated space heating zone of the storage and fed to the space heat distribution. Thus, heat that was delivered by the heat pump with flow temperatures of 40–50 °C is subsequently mixed down in the space heat flow control mixer to 30 °C needed for space heating. It should go without saying that running a heat pump at sink temperatures that are 10–20 K higher than needed is detrimental for its performance.

5.3.2.2 Exergetic losses in general

Figure 5.8 shows the accumulated energy (heat) that was measured during a 12-day test below a certain supply temperature that is displayed on the x-axis. It can be seen that the HP delivered its heat for space heating ($Q_{HP,SH}$) during this test at a temperature level that is around 8 K higher than the temperature level of the space heat distribution (Q_{SH}). Assuming that the COP of the heat pump decreases by about 2%/K, an increase in electric energy demand for the heat pump of approximately 16% can be expected, corresponding to a reduction of more than 0.5 points of the performance factor.

5.3.3 Extension of DST test procedure toward solar and heat pump systems

Development of a new test procedure was initiated due to lack of an appropriate test procedure to characterize the performance of heat pump systems for domestic hot water preparation, which use unglazed collectors to evaporate the refrigerant (instead of an external air unit with a fan). Such systems cannot be characterized with CTSS, for example, because the phase change processes cannot be characterized with the collector test procedures described in ISO 9806. Looking at heat pump test standard EN 16147:2011 could be considered. However, the standard does not allow solar assistance in the evaporator. Furthermore, the system is only tested at a single ambient air temperature.

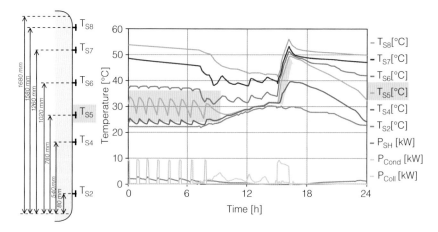

Fig. 5.7 Measured temperatures and power on the 5th test day of a 12-day test. T_{S8}–T_{S2} = temperature measured with contact sensors on the tank; P_{SH} = power of space heating; P_{Cond} = power supplied by condenser; P_{Coll} = power supplied by solar collectors

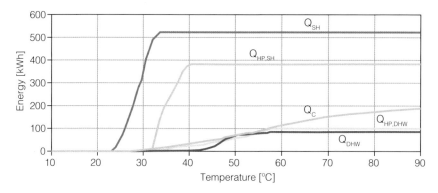

Fig. 5.8 Measured data of the accumulated energy over the temperature where it was supplied ($Q_{HP,SH}$, $Q_{HP,DHW}$, and Q_C) or consumed (Q_{DHW} and Q_{SH}) for a 12-day test. The decisive temperature in each case was the flow temperature

Within the course of the development activities, such a system was installed at the test rig with the evaporator unit (collector) outdoor and measured in Lisbon during 1 year covering almost all local possible weather conditions. The hot water tapping test cycle used is in agreement with the standard EN 16147:2011. Exceptions were, as indicated earlier, variable outside air temperature and additionally solar irradiance on the evaporator. The heat storage tank of the system had a rated volume of 300 l and an evaporator surface area of 1.6 m². The setpoint temperature for the heat pump was set to 50 °C. EN 16147:2011 is based on EU reference tapping cycles. According to ecodesign requirements [28], the maximum tapping cycle or the tapping cycle one below the maximum tapping cycle shall be applied. Tapping test cycle "XL" was tested but the system was unable to achieve an average outlet temperature of 40 °C at 21 : 30 (tapping

5.3 Experience from laboratory testing

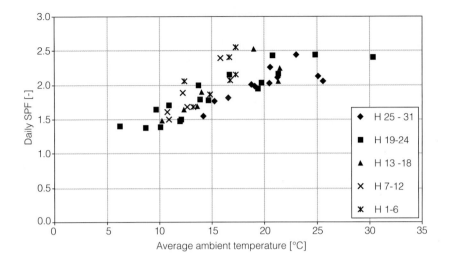

Fig. 5.9 Daily SPF as a function of daily solar irradiation and average daily air temperature

test type bath). Tapping cycle "L" was chosen instead, as the system was able to satisfy the requirements.

The performance of the system is presented in Figure 5.9 as a function of outdoor conditions: daily average ambient air temperature and daily solar irradiation. The thermal performance of the heat pump shows a correlated dependence with the daily average ambient air temperature. The influence of solar irradiation on overall system performance was smaller than that expected: A dependence of the daily solar irradiation with higher SPF (seasonal performance factor) values cannot be observed. The main reason can be found in the load profile of the tapping cycle. Most of the energy is extracted in the morning and at the evening, forcing the compressor to operate predominantly in periods with low irradiance.

Along with the vast amount of measurement data, a short-term laboratory test method could be developed. Four specific days were chosen: Two cold days – one with high- and one with low-irradiation conditions – and two warm days with irradiation conditions analogous to the cold day. The measurement data were used to fit a model by means of parameter identification in order to generate long-term performance figures (modeling and simulation test data evaluation method with "close to reality" test sequences, cf. Figure 5.3).

Thus, a numerical model was developed in TRNSYS based on the model by Morrison [29]. For the model of the storage tank, however, a modification of Type 4, instead of Type 38 was applied, because the heating rate in the condenser inside the storage is variable in Type 4. The heat pump evaporator is simulated as an uncovered solar collector. The influence of condensation when the collector temperature is below the dew point temperature is considered. Measurement data were then fitted with certain parameters of the model in order to adjust the model so that it can reproduce the thermal performance of the investigated system.[3]

---

[3] More details on the used equations and identified parameters can be found in Ref. [14].

Table 5.6 Experimental validation of the developed model

	Test day	T_{air} (°C)	G (MJ/m^2)	$T_{air,int}$ (°C)	Exp. comp. energy (kWh)	Model. comp. energy (kWh)	Relative error (%)
Cold/high irradiation	February 3, 2012	6.2	22.9	10.2	8.49	8.91	5.0
Cold/low irradiation	January 25, 2012	14.7	7.4	12.8	7.58	7.42	2.1
Warm/high irradiation	June 24, 2012	30.3	22.2	23.2	4.69	4.73	0.9
Warm/low irradiation	June 14, 2014	21.3	18.0	21.3	5.58	5.55	0.6

Table 5.6 presents experimental validation of the model for four specific days. The maximum relative error in objective function – electrical compressor energy consumption – was about 5.0%.

Further steps regarding the methodology include testing of more solar-assisted heat pump systems for DHW with different configurations (different storage volume and evaporator area), model improvement in order to increase accuracy, and an implementation of a software capable of calculating the annual thermal performance for different locations. The aim is to frame this kind of systems as "renewable energy equipment" in current Portuguese regulation and to implement a procedure capable of estimating the seasonal performance factor. According to the European Directive 2009/28/EC [30], only heat pumps with SPF > 1.15/η (with η being the ratio between total gross production of electricity and primary energy consumption for electricity production) should be considered as renewable energy equipment.

5.4 Summary and findings

Test procedures used to characterize the performance for SHP systems have been recently developed, based on test procedures to a large extent originating from the field of solar thermal applications. Considering the number of market-available systems identified during the T44A38 market survey, more than 70% of the system configurations found can be characterized with the test methods on current level of development at the end of 2013 (cf. Section 5.2.3). Performance test procedures for SHP systems can be assigned to two main approaches: component testing system simulation and whole system testing.

Component-wise testing (CTSS) is suitable for component manufacturers and SHP system dealers, who can assess the performance of custom-designed systems under different boundary conditions and for labeling purposes (energy labeling also for SHP systems is mandatory from September 2015 in the European Union). The performance-relevant components of SHP systems, collector, thermal storage tank(s), heat pump, and system controller are characterized individually. Performance evaluation of the system is carried out by simulation and the parameterization of the system models is based on the results of the tested components. In principle, performance can be assessed for various climates and user load patterns. This approach can significantly reduce cost of testing of a larger product portfolio by allowing identical components employed in several systems to be tested only once.

In integrated systems, for example, compact packages of a heat pump with a storage, or for complex control, it might be difficult for this test method to predict the overall system performance, especially when considering all aspects of component integration and their interactions. Here, "whole system testing" is the more useful approach. In contrast to component-wise testing, all parts are installed at the test rig as delivered by the manufacturer. The strength of the whole system test is a more profound performance characterization compared with component testing by considering system interactions, close to reality test sequences, and autonomous controller operation during the test. Performance calculation is done either by direct extrapolation of the performance figures measured during the test or by modeling and simulation. For compact systems, this is likely to be also the more cost-effective approach, since only one test is needed instead of one for each component.

A drawback of all test procedures relying on modeling and simulation is that the characterization of a new developed component may be a problem if no proper model is available. Furthermore, it must be noted that models and the determination of model parameters introduce high uncertainties, in particular for new products, new models, or unusual operating conditions in new system concepts. Perfect assumptions are often applied for simulation, which can be naive to assume, especially when newly developed products are brought into market. In the end, the cost of a test that is not mandatory by regulation should oppose to the cost occurring if failures are detected later with the customer. Experience has shown that electricity consumption of badly configured SHP systems can be decreased by 50% – for identical components, and even good systems could be still improved [20,21].

It can be observed that manufacturers initiate time-consuming field testing in order to prove the performance of their system. Often, as systems are underperforming, additional measures are taken in the course of this process. The effectiveness of such measures, however, is less often clear to evaluate, because the measurement equipment might be inaccurate or not located properly in order to calculate energy balances. Furthermore, results from field tests cannot be compared with each other, since the performance results may be determined more by the climate, the heat loads, and the user behavior, than by the quality of the SHP system. Such problems can be overcome by carrying out "accelerated emulated field testing" by means of a whole system test (cf. Section 5.3.2). Apparently, it is more cost efficient to check proper system function and identify the optimization potential in a test laboratory with more high-quality measurement equipment and experienced staff. In the second step, a field test creates confidence in long-term reliable operation.

Standards are available for the component-wise testing approach; however, SHP systems are not yet included. Considering the whole system test procedures for combined systems, harmonization of a number of test procedures is currently ongoing. These activities may also lead to standardization. Said test procedures are on a high level of development, experience has been gained with the testing of numerous types of combined solar thermal systems. Especially for SHP systems where the collector acts as the evaporator of the heat pump, new whole system test procedures have been developed (cf. Table 5.2, related test procedures, Section 5.3.3).

SHP systems are also covered in the EU Ecodesign and Energy Labeling Directive L239 [31]. SHP systems can be labeled with three different types of "package labels" (domestic hot water only, space heating only, or combined systems). Since SHP systems incorporate the highest fraction of renewable energies, they can be expected to be labeled between A+ and A+++. The basis for the energy label is a calculation based on design parameters of the heating system, which have to be determined from component tests. The logic behind the simple calculation procedure to determine the efficiency class for the package label could be expressed like "the better the components (i.e., rated power of the heater (↓), collector area (↑) and efficiency (↑), thermal storage volume (↑) and heat loss rate (↓), control class (↑)), the better the primary energy efficiency of the overall SHP system." On the other hand, experience gained during T44A38 from monitoring projects and whole system testing has shown that a system constructed out of good components does not at all allow the conclusion that overall system performance is equally good.

References

1. Mahlia, T.M.I. (2004) Methodology for predicting market transformation due to implementation of energy efficiency standards and labels. *Energy Conversion and Management*, **45** (11–12), 1785–1793.

2. Meier, A.K. and Hill, J.E. (1997) Energy test procedures for appliances. *Energy and Buildings*, **26** (1), 23–33.

3. Malenkovic, I. and Serrats, M. (2012) Review on Testing and Rating Procedures for Solar Thermal and Heat Pump Systems and Components, QAiST Technical Report 5.1.2.

4. Haberl, R., Haller, M.Y., Bales, C., Persson, T., Papillon, P., Chèze, D., and Matuska, T. (2012) Dynamic whole system test methods – overview and current developments. Presentation held at the IEA SHC Task 44/HPP Annex 38 "Solar and Heat Pump Systems" Experts Meeting at Póvoa de Varzim (Porto), Portugal.

5. EN (2006) EN 12975-2:2006 – Thermal Solar Systems and Components. Solar Collectors. Part 2. Test Methods.

6. ISO/TC 180 (2013) ISO 9806:2013 – Solar energy. Solar thermal collectors. Test methods.

7. CEN (2012) EN 12977-3:2012 – Thermal solar systems and components. Custom built systems. Part 3. Performance test methods for solar water heater stores (German version).

8. CEN (2012) EN 12977-4:2012 – Thermal solar systems and components. Custom built systems. Part 4. Performance test methods for solar combistores (German version).

9. CEN (2012) EN 12977-5:2012 – Thermal solar systems and components. Custom built systems. Part 5. Performance test methods for control equipment (German version).

10. Haller, M.Y., Haberl, R., Persson, T., Bales, C., Kovacs, P., Chèze, D., and Papillon, P. (2013) Dynamic whole system testing of combined renewable heating systems – the current state of the art. *Energy and Buildings*, **66**, 667–677.

11. ISO/TC 180 (2013) ISO 9459-5:2007 – Solar heating. Domestic water heating systems. Part 5. System performance characterization by means of whole-system tests and computer simulation.

12. Mette, B., Drück, H., Bachmann, S., and Müller-Steinhagen, H. (2009) Performance testing of solar thermal systems combined with heat pumps. Solar World Congress 2009, Johannesburg, pp. 301–310.

13. Panaras, G., Mathioulakis, E., and Belessiotis, V. (2014) A method for the dynamic testing and evaluation of the performance of combined solar thermal heat pump hot water systems. *Applied Energy*, **114**, 124–134.

14. Facão, J. and Carvalho, M.J. (2014) New test methodologies to analyse direct expansion solar assisted heat pumps for domestic hot water. *Solar Energy*, **100**, 66–75.

15. Vogelsanger, P. (2002) The Concise Cycle Test Method – A Twelve Day System Test. A Report of IEA SHC – Task 26. International Energy Agency Solar Heating and Cooling Programme.

16. Bales, C. (2004) Combitest – a new test method for thermal stores used in solar combisystems. Ph.D. thesis, Building Services Engineering, Department of Building Technology, Chalmers University of Technology, Gothenburg, Sweden.

17. Leconte, A., Achard, G., and Papillon, P. (2012) Global approach test improvement using a neural network model identification to characterise solar combisystem performances. *Solar Energy*, **86** (7), 2001–2016.

18. Walter, É. and Pronzato, L. (1997) *Identification of Parametric Models from Experimental Data*, Springer.

19. Drück, H. (2006) Multiport Store Model for TRNSYS – Type 340 – V1.99F.

20. Haberl, R., Haller, M.Y., Reber, A., and Frank, E. (2014) Combining heat pumps with combistores: detailed measurements reveal demand for optimization. *Energy Procedia*, **48**, 361–369.

21. Haller, M.Y., Haberl, R., Mojic, I., and Frank, E. (2014). Hydraulic integration and control of heat pump and combi-storage: same components, big differences. *Energy Procedia*, **48**, 571–580.

22. Bachmann, S., Drück, H., and Müller-Steinhagen, H. (2008) Solar thermal systems combined with heat pumps – investigation of different combisystem concepts. Proceedings of the EuroSun 2008 Conference, Lisbon.

23. Universität Stuttgart, WPSol – Leistungsprüfung und ökologische Bewertung von kombinierten Solar-Wärmepumpenanlagen, Institut für Thermodynamik und Wärmetechnik, Universität Stuttgart. Available at http://www.itw.uni-stuttgart.de/forschung/projekte/aktuell/wpsol.html (accessed June 11, 2014).

24. Haberl, R., Frank, E., and Vogelsanger, P. (2009) Holistic system testing – 10 years of concise cycle testing. Solar World Congress 2009, Johannesburg, South Africa, pp. 351–360.

25. Papillon, P., Albaric, M., Haller, M., Haberl, R., Persson, T., Pettersson, U., Frank, E., and Bales, C. (2011) Whole system testing: the efficient way to test and improve solar combisystems performance and quality. ESTEC 2011 – 5th European Solar Thermal Energy Conference, October 20–21, Marseille, France.

26. Haller, M.Y., Haberl, R., Carbonell, D., Philippen, D., and Frank, E. (2014) SOL-HEAP – Solar and Heat Pump Combisystems, im Auftrag des Bundesamt für Energie BFE, Bern.

27. Haberl, R., Haller, M.Y., and Frank, E. (2013) Combining heat pumps with combistores: detailed measurements reveal demand for optimization. SHC Conference 2013, Freiburg, Germany.

28. European Commission (2013) Commission Regulation (EU) No 814/2013 of 2 August 2013 implementing Directive 2009/125/EC of the European Parliament and of the Council with regard to ecodesign requirements for water heaters and hot water storage tanks. *Official Journal of the European Union*, **L 239**, p. 162–183.

29. Morrison, G.L. (1994) Simulation of packaged solar heat-pump water heaters. *Solar Energy*, **53** (3), 249–257.

30. The European Parliament and the Council of the European Union (2009) Directive 2009/28/EC of the European Parliament and of the Council on the promotion of the use of energy from renewable sources and amending and subsequently repealing Directives 2001/77/EC and 2003/30/EC. *Official Journal of the European Union*, **L 140**, p. 16–62.

31. The European Parliament and the Council of the European Union (2013) *Official Journal of the European Union*, doi:10.3000/19770677.L_2013.239.eng see: http://eur-lex.europa.eu/legal-content/EN/TXT/?uri=OJ:L:2013:239:TOC **L 239/56**.

Part Two
Practical Considerations

Solar and Heat Pump Systems for Residential Buildings, First Edition.
Edited by Jean-Christophe Hadorn.
© 2015 Ernst & Sohn GmbH & Co. KG. Published 2015 by Ernst & Sohn GmbH & Co. KG.

6 Monitoring

Sebastian Herkel, Jörn Ruschenburg, Anja Loose, Erik Bertram, Carolina de Sousa Fraga, Michel Y. Haller, Marek Miara, and Bernard Thissen

Summary

The main aims of the *in situ* monitoring are the evaluation of performance of solar and heat pump (SHP) systems under real-life conditions, the detection of installation errors, the optimization of the entire systems' operation as well as control functions for different operation modes, and finally the dimensioning of the collector field and storage capacity. *In situ* monitoring of more than 40 SHP installations (field tests) was performed in IEA T44A38 and is presented in this chapter. *In situ* monitoring needs both a well-defined procedure and an accurate measurement and data acquisition technology. Clear boundary conditions and balance borders are needed for a common evaluation. The seasonal performance factor of the entire SHP system (SFP_{SHP}) including all thermal storages as presented in Chapter 4 has been found to be appropriate.

The participants from seven countries have been able to provide 1–2 years of monitored results. Results showed the marketed variety of systems, though some monitored systems still have to be considered as prototypes. The variance of results was found to be large with observed SPF_{SHP} ranging from 1.5 to 6 with a median value of 3.0. Reasons for the variety of results have been analyzed. In general, a good quality assurance during design and installation was found as key for a good performance. The size and quality of the storage and low system temperatures for heating distribution have been identified as key indicators. When neglecting the storage losses and some of the auxiliary pumps, the system performance SPF_{bSt} was 4.1 – a value that is rather high when compared with other field tests with heat pump only systems. Although parallel systems are the most common and the simplest to operate, good performing systems were found in all categories of hydraulic layouts and good integration of all components was shown to be possible. Some best practice examples are presented in greater detail.

6.1 Background

This chapter describes the *in situ* monitoring (so-called field test) of combined solar thermal and heat pump systems, that is, the metrological attendance and examination of real heating systems in different countries over the period of at least 1 year.

Broad field tests of the separate system technologies have been performed in the past already for heat pump systems only and for solar thermal and storage systems without heat pumps [1], but not yet for the specific combination of solar thermal and heat pump systems. Although some of these combined systems have been monitored in single cases, a systematic study related to the *in situ* performance for this system category is still missing. Therefore, field tests based on *in situ* monitoring have been performed in international cooperation within IEA SHC Task 44/HPP Annex 38 in order to determine the thermal performance of combined solar thermal and heat pump systems under real operating conditions.

The main aim of the *in situ* monitoring is, on the one hand, the detection of installation errors, further development of components, optimization of the operation behavior of the entire systems and controlling functions for different operation modes, and the dimensioning of the collector field and storage capacity. On the other hand, measured data are needed for the validation of numerical simulation models of combined solar thermal and heat pump systems and components.

In addition to results from laboratory testing procedures, it is therefore crucial for these aims to have measured data gained from field tests of real installations. Also, the systems' thermal performance can be compared with theoretical predictions, for example, from simulations. However, it has to be underlined at this point that mere field test results from *in situ* monitoring cannot be directly compared, since the results measured depend on a variety of boundary conditions such as location of the building, climatic conditions, building efficiency standard and space heating demand of the building, and last but not least the user behavior. Even the very same system in the same location will result in different performance figures when measured in different years, depending on the boundary conditions, for example, if the winter is mild or cold. Due to this fact, only with the help of simulations of the systems under reference conditions the results of different SHP system configurations become directly comparable.

6.2   Monitoring technique

6.2.1   Monitoring approach

Within T44A38, the approaches of the monitoring procedure for field test objects were very similar for different research institutes and will be treated within this chapter in a generalized way.

New performance factors and system boundaries for combined SHP systems have been agreed on within the task. They have already been described in Chapter 4 and are used for the monitoring procedure and data analysis as well. As can be derived from the large variety of different boundaries and possible SPF values, there are many ways to monitor an SHP system, ranging from very simple to rather complicated and detailed. Therefore, depending on the aim of the measurements the complexity of the monitoring equipment and resulting from this also its costs can vary a lot. The most easiest way to monitor an SHP system would be to measure only the amount of produced useful heat for domestic hot water (DHW) preparation and space heating on the one hand, and the total electric energy consumption of the overall system on the other hand, for instance, on a monthly basis only. By measuring these two values, the performance factor SPF_{SHP} or SPF_{SHP+} (which includes additionally the electricity for the heating distribution pump) can already be determined in order to have a first hint for comparison of the systems' performances. However, much more information can be derived from a more detailed measurement strategy. Depending on the main goal or interest of the investigation,

6.2 Monitoring technique

different questions may be addressed. The performance factor SPF_{bSt}, for example, is very useful for the comparison of an SHP system with conventional heating systems, foe example, gas boilers, as it does not consider storage losses. This performance figure is often used in field test for heat pumps only [1]. The determination of the SPF of single components as the heat pump only (SPF_{HP}) might be useful for comparison with monovalent heat pumps. Performances of single components are needed also for the development of component-oriented laboratory test procedures, that is, for the validation of mathematical models for the thermal behavior of the heat pump or other components under dynamic operating conditions. For any performance figure to be evaluated, all energy flows have to be measured that cross the boundary corresponding to the respective performance figure's definition.

In Figure 6.1, an exemplary parallel system is depicted, based on the visualization method presented in Chapter 2, but with possible detailed monitoring equipment. Here, the blue cycles labeled with "P_{el}" depict electricity meters and the blue squares are heat meters. In addition to these types of meters, also temperature sensors alone might be installed for special questions, for example, for the monitoring of the thermal stratification of combi-storages or in the refrigerant cycle of an air/water split heat pump, where

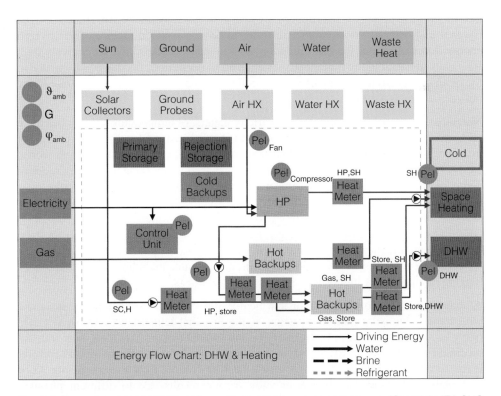

Fig. 6.1 Exemplary parallel SHP system with monitoring equipment shown. (Copyright IEA SHC Task 44/HPP Annex 38.)

no flow meter can be installed. Defrosting of air/water heat pumps can also be monitored though this operation mode is not shown in the figure.

In the ideal case, a heat meter is installed in each circuit depicted as a red arrow in Figure 6.1, because it is also of interest to be able to distinguish between different possible operational modes of the SHP system. If, for instance, the heat pump may be used for direct space heating or for charging a heat store, it makes sense to distinguish between these two quantities. However, in real monitoring situations, heat flows will often be merged into a single value, for example, for space heating including different heat sources. If a storage charging circuit might be used for solar defrosting or heating out of the heat store, a heat meter with heating and cooling functions – or volume flow meter that can also detect reverse flow directions – must be installed. In case of an installed domestic hot water circulation circuit, also here the heat flow and the electricity consumed by the circulation pump are measured in order to determine the amount of heat losses and the additional electricity consumption due to this circulation modus. Also the ratio of different heat generators' heating (e.g., solar thermal, geothermal, and backup) is often of interest. In this case, the heat flows of the generators have to be measured separately.

For model-based evaluation of complete systems, that is, by means of simulation, or for the validation of numerical component models, a more complex and highly differentiated monitoring strategy is required. Many heat flows and electricity consumers are to be covered and determined separately in order to have a detailed picture of the overall energy flows. As input for the validation of simulation models, also a relatively high frequency of data logging time steps is crucial. Most institutions that have participated in the monitoring activities of T44A38 have collected data as mean values in a time frame between every 1–5 min.

Not only the number of heat meters, electricity meters, or temperature sensors has to be taken into account, but the exact positioning is also a decisive factor in some cases. For instance, if the comparison of different components (types of heat pumps, heat stores, etc.) is of interest, the sensors need to be always placed in the same position in order to count for distribution losses in the same way. Another detail that must not be neglected is the accuracy and resolution of the different measurement devices. It has to be taken into account that small volume flows (e.g., for domestic hot water tapping) or electricity consumers with very low power (e.g., the controller or circulation pumps) must be detectable and the measurement equipment must be selected in an appropriate way. An uncertainty estimation of the whole measurement chain, including the resolution of logging equipment, is highly recommended in field tests.

A characteristic of combined solar and heat pump systems, compared, for instance, with stand-alone solar thermal systems, is the fact that temperatures below 0 °C are likely to appear, namely, in the primary circuit of heat pumps. Also, the heat transfer medium (brine) is often different from those used in solar thermal systems. Therefore, it has to be taken into account that many heat or volume flow meters do not work accurately under such conditions or show wrong results due to different physical properties of the medium compared with water (e.g., density, specific heat capacity, or viscosity). Another feature of SHP systems is the fact that solar thermal collectors are being operated under lower operating temperatures than usually, especially when integrated in a series system (cf. Chapter 2). In particular, unglazed solar collectors or absorbers are affected by this phenomenon and in some cases frozen temperature sensors have been seen, which is not supposed to occur in a conventional solar thermal system. It is also helpful to measure the relative humidity and temperature of the ambient air in order to be able to investigate further frosting and defrosting phenomena of air/water heat pumps.

6.2.2 Measurement technology

This section gives a brief overview about the technical equipment used for the monitoring of SHP systems in a generalized way. The most important components are a data logger, heat flow meters, electricity meters, temperature sensors, and some equipment for measurement of the ambient conditions such as ambient temperature, relative humidity, or solar irradiation.

6.2.2.1 Data logging systems

A variety of different data loggers have been used and can be looked up in the reports of the respective institutions. Most institutions used a remote control and daily to weekly data transfer via Internet or mobile communications using GSM modems.

The measurement intervals of data points were rather short, and vary between 1 and 30 s depending on the type of sensor. An average or sum of these measured values was then stored by the data logger in intervals between 1 and 5 min.

Due to this high frequency of monitoring, a huge database of raw data needs to be handled, checked for plausibility, and then analyzed. This procedure has been dealt with in very different ways by the institutes so that at this point further discussion is refrained from.

Figure 6.2 shows an example for a data logging system that uses a modem for communication and data transfer over long distances.

6.2.2.2 Heat meters

For the determination of heat flows, the volume or mass flow rate in the examined circuit needs to be measured along with the respective flow and return temperatures. From these measured values, the heat flow can be calculated applying the temperature-dependent specific heat capacity and the density (in case of measured volume flow rate) as physical properties of the heat transfer medium. Some heat meters feature an own integrated arithmetic unit that calculates the values from the data measured. This is especially the case for water as heat transfer medium. In all other cases, the heat flow has to be calculated from the raw data set.

Ultrasonic volume flow meters with a calibrated pair of Pt 500 or Pt 1000 temperature sensors were often used for monitoring the heat flow in water-filled circuits or in solar circuits. Fluid temperatures are measured by the use of diving sensors (sensor placed in a tube reaching into the fluid stream, in most cases) or Pt 1000 probes clamped onto pipes

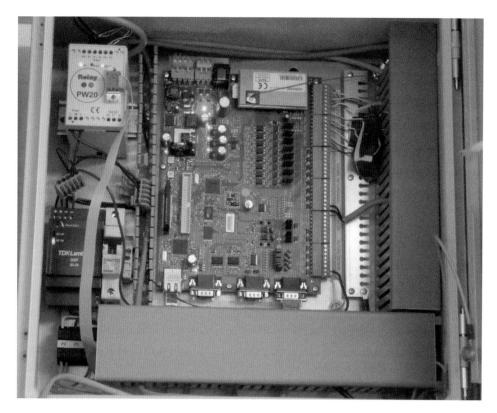

Fig. 6.2 Exemplary data logging system with input signal via digital, analog, and temperature channels or an MBUS connection [2]

underneath pipe insulation with heat conductive paste. For circuits within the system that do contain other heat transfer media, for example, mixtures based on ethylene glycol and water with different concentration ratios, ultrasonic heat meters are not recommendable due to the liquids' deviating viscosities, especially at low temperatures that might occur in the brine loops of a heat pump. In this case, simpler turbine-type volume flow meters with pulses as output are used, for instance. However, at temperatures below 0 °C, these are no more reliable.

6.2.2.3 Electricity meters

Electricity meters have been used for monitoring the electricity consumption of the heat pump itself (compressor), the control unit, the electric heating element if foreseen as backup, all circulation pumps, control unit of the solar thermal circuit if applicable, and the fan for air source heat pumps. Most electricity meters were equipped with a mechanical counter as well as an impulse interface with resolutions between 100 and 2000 pulses per kWh, depending on the consumer. For the heat pump and the electric heating elements, three-phase current meters were necessary, while for all other consumers 230 V one-phase meters were used.

6.3 Solar and heat pump performance – results from field tests

6.2.2.4 Meteorological data

It is not implicitly necessary to have measured data of the solar irradiation within the monitoring procedure. However, most institutes have recorded also the solar irradiation onto the inclined collector plane, using equipment ranging from rather simple and cheaper silicon cells to high-precision pyranometers. Another meteorological parameter frequently monitored is the ambient temperature. Also, the relative humidity of the ambient air surrounding air source heat pumps has been measured in many cases. For more special applications (e.g., uncovered solar collectors), additional parameters have been recorded, for example, long-wave radiation, wind velocity, or soil temperatures.

6.2.2.5 Temperature sensors

Apart from the ambient temperature and heat meters already mentioned above, single temperature sensors have found application in many cases as well. Some examples are the measurement of the thermal stratification of combi-storages (sensors placed directly on the surface of the storage beneath the insulation at defined heights) and the temperatures in the primary loop of air/water heat pumps (refrigerant cycle) where no volume flow meter can be placed, or in different depths within a borehole heat exchanger in order to observe, for instance, regeneration processes of the earth.

Temperature sensors were in most cases Pt 100 or Pt 1000 sensors with different classes of accuracy. For the use in heat meters, they were often calibrated not in absolute way but pairwise to increase the accuracy of the measurement. Four-wire and two-wire temperature sensors have been applied.

6.3 Solar and heat pump performance – results from field tests

The purpose of *in situ* measurements of SHP systems under real conditions is to better understand the interoperation of different components of these systems and to evaluate the performance of these systems. Within the framework of T44A38, members of 18 research institutions monitored 45 systems in the last few years; those selected for further evaluation are listed in Table 6.1. The solar and heat pump systems analyzed

Table 6.1 Characteristics of the monitored systems

Project name or description		Organization	Concept	Collector type	HP source
Solpumpeff project A	Austria	AEE Intec/TU Graz	P/S	FPC	Air/solar
Solpumpeff project B	Austria	AEE Intec/TU Graz	P	FPC	Air
Solpumpeff project C	Austria	AEE Intec/TU Graz	P	FPC	Ground
Solpumpeff project D	Austria	AEE Intec/TU Graz	P	FPC	Air
Solpumpeff project E	Austria	AEE Intec/TU Graz	P/S	FPC	Air/solar
Solpumpeff project F	Austria	AEE Intec/TU Graz	P	FPC	Water

(continued)

Table 6.1 (*Continued*)

Project name or description		Organization	Concept	Collector type	HP source
Savièse Aufdereggen ice storage	Switzerland	Energie Solaire SA	P/S	UC	Solar
Savièse Granois ice storage	Switzerland	Energie Solaire SA	P/S	UC	Solar
Fribourg	Switzerland	EIA Fribourg	P/S	FPC	Ground
Jona	Switzerland	Institut für Solartechnik SPF	P	FPC	Air
COP5	Switzerland	Université de Genève	P/S	UC	Solar
WP Effizienz A	Germany	Fraunhofer ISE	P	ETC	Air
WP Effizienz B	Germany	Fraunhofer ISE	P	FPC	Air
WP Effizienz C	Germany	Fraunhofer ISE	P	FPC	Ground
WP Effizienz D	Germany	Fraunhofer ISE	P (DHW)	FPC	Air
WP Effizienz E	Germany	Fraunhofer ISE	P (DHW)	FPC	Ground
WP Effizienz F	Germany	Fraunhofer ISE	P (DHW)	FPC	Air
WP Monitor A	Germany	Fraunhofer ISE	P (DHW)	FPC	Ground
WP Monitor B	Germany	Fraunhofer ISE	P	FPC	Ground
WP Monitor C	Germany	Fraunhofer ISE	P (DHW)	ETC	Ground
Haus der Zukunft	Germany	Fraunhofer ISE	P/S	FPC	Air/solar
Dreieich	Germany	ISFH	R	PVT	Ground
Limburg	Germany	ISFH	R	UC	Ground
Solar and heat pump for houses A	Germany	ITW	P	FPC	Air
Solar and heat pump for houses B	Germany	ITW	P	FPC	Air
Solar ice store	Germany	ITW	P/S	FPC/UC	Solar
Borehole HX	Germany	ITW	P/S	FPC	Ground/solar
Solar and heat pump for houses C	Germany	ITW	P	FPC	Air
Energy basket and unglazed absorber	Germany	ITW	P/S/R	UC	Ground/solar
Balleruphuset	Denmark	Cenergia/SBI	P/S/R	FPC	Ground/air
Flamingohuset	Denmark	Cenergia/SBI	P/R	FPC	Ground
Heliopac	France	EDF	S	UC	Solar

Abbreviations: FPC = flat-plate collector; UC = uncovered collector; ETC = evacuated tube collector; PVT = photovoltaic–thermal collector.

6.3 Solar and heat pump performance – results from field tests

varied in terms of system layout, size, location, and underlying research questions leading to different measurement concepts and monitoring equipment used. Like all activities of T44A38, the monitored SHP systems and the subsequent analyses are limited to systems that are equipped with electrically driven compression heat pumps and designed for DHW preparation and/or residential space heating. The system analysis is presented anonymously, giving each of the systems a number code.

Out of 32 analyzed systems, 17 could be characterized according to the systematic introduced in Chapter 2 as parallel systems (P), 9 as parallel/series systems (P/S), and 5 as regeneration of the ground by uncovered or PVT collectors (R or P/S/R). Thirteen systems use ambient air as main source for the heat pump, 14 systems use the ground or water, and 5 systems exclusively use the solar absorber.

The nominal coefficient of performance (COP) of the installed heat pumps under standard test conditions varies from 2.9 to 4.2 for air/water systems and from 3.8 to 4.7 for brine/water systems and reflects the bandwidth of actual marketed products. The different testing standards EN 255 and EN 14511 should be recognized as the latter, newer procedure leads to lower performances. The variation of the COP of air source systems is higher than that of the ground source systems (Figure 6.3).

A key indicator of the performance of a system is the seasonal performance factor (for definitions see Chapter 4). Due to different underlying research questions, the measurement equipment and the location of the meters vary in the analyzed systems as described in Section 6.1. Consequently, the chosen boundary for determining the SPF varies in the projects. In order to compare the system's SPF, the measured indicators were normalized to the boundary including storage losses and loading pumps, that is, SPF_{SHP}. The performance difference between the two boundaries SPF_{bSt} and SPF_{SHP} strongly depends on the ratio of heating demand to storage losses and therefore on the size of the storage. Based on the results of 13 field tests where both figures were available, a

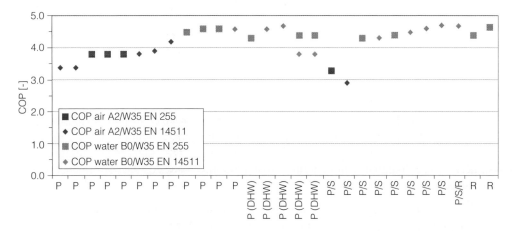

Fig. 6.3 Nominal COP of the installed heat pumps. On the abscissa, the type of SHP system is given. The shape of the marker refers to the standard used for testing; the color refers to the kind of heat pump

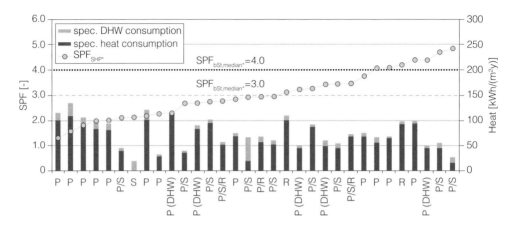

Fig. 6.4 Measured or correlated SPF^*_{SHP} of the monitored systems (green markers, left ordinate) and heat consumption for domestic hot water and space heating (right ordinate). The type of system is given on the abscissa: P = parallel; S = series; R = regenerative (cf. Section 2.1.3)

correlation was used to derive the SPF_{SHP} based on SPF_{bSt} and marked with an asterisk for distinction:

$$SPF^*_{SHP} = 0.75 \cdot SPF_{bSt} - 0.22$$

Similarly, for SPF_{SHP+}, a simplified correlation out of 15 installations is found as

$$SPF^*_{SHP} = SPF_{SHP+} + 0.15 \quad \text{(to be added as constant value)}.$$

Figure 6.4 gives the range of measured SPF^*_{SHP} for the different field tests. In addition, the heat consumption for domestic hot water and room heating related to the floor is given. The SPF^*_{SHP} varies from 1.3 to 4.8 with a mean value of 3.02. When the storage losses are not taken into account, the mean SPF_{bSt} is 3.95. Thus, the storage losses in SHP systems have a strong impact on the performance.

The space heat consumption varied from 15 to 110 kWh/(m² year) with a median value of 67 kWh/(m² year) and domestic hot water consumption between 4 and 48 kWh/(m² year) with a median value of 11 kWh/(m² year). The share of domestic hot water varies between 6 and 100%. So, on average standard low-energy houses with a relatively low share of DHW were analyzed. Compared with the common heat pump field test, the average SPF_{bSt} is comparable to average ground source heat pump systems. Reflecting the fact that some of the tested systems were prototypes, the results can be interpreted as success even though the strong influence of the storage losses on the performance shows potential for improvement.

In order to deeply analyze the monitored systems, these were grouped by their main source and the influence of key indicators on the performance was identified. It can be expected that an increasing size of the absorber in relation to the installed heat pump capacity leads to a higher SPF, as the solar fraction will increase, and a higher solar fraction implies a higher SPF. A second key parameter is the COP of the heat pump

6.3 Solar and heat pump performance – results from field tests

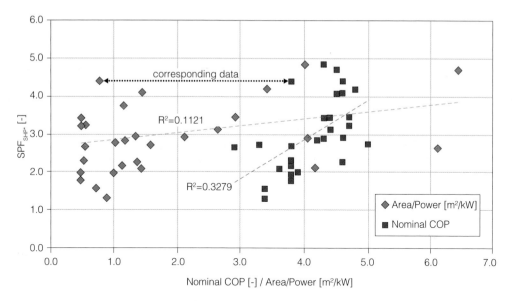

Fig. 6.5 Measured or correlated SPF^*_{SHP} depending on the nominal COP (A2/W35 or B0/W35) and the ratio of installed absorber area and nominal heat pump capacity

under standard conditions. A third set of parameters includes mean temperatures at the evaporator and condenser of the heat pump. As for the latter one data were not available, the ratio between domestic hot water demand and the heating demand could be used instead, giving an indication for the temperatures on the condenser side (the lower the share, the lower the temperatures and the higher the performance).

In Figure 6.5, the measured SPF^*_{SHP} is shown as a function of nominal COP and the ratio of installed absorber versus heat pump capacity. Selected systems show the potential of air and solar only based SHP systems, and they can achieve SPF^*_{SHP} significantly higher than 4. The results for solar ground source heat pumps are as widespread as for the air source heat pumps. These systems achieved higher SPF^*_{SHP} in general; best systems achieved a SPF^*_{SHP} of 4.8. Nevertheless, the rather weak performance of many systems underlines the necessity of a well-established quality assurance process for installation, especially when hydraulics and control become more complex – even though many of the monitored systems had an experimental character.

As expected, the system performance gets lower as the share of high-temperature DHW increases (Figure 6.6). The dependence was found to be strongest for air source heat pumps. Especially, the three series systems with a high share of DHW and uncovered collectors as only source show this tendency. The installed collector area in relation to the installed heat pump capacity ranges from 0.5 to 1.5 m^2/kW for flat-plate collectors. Larger areas were installed when uncovered collectors were used.

Figure 6.7 shows the dependence of the systems on the ratio of installed aperture area per heat consumption. Larger solar installations lead to higher performances as expected. Apart from appropriate sizing of the collector subsystem, there must be other influences on the SPF^*_{SHP} performance values for air source heat pumps, for

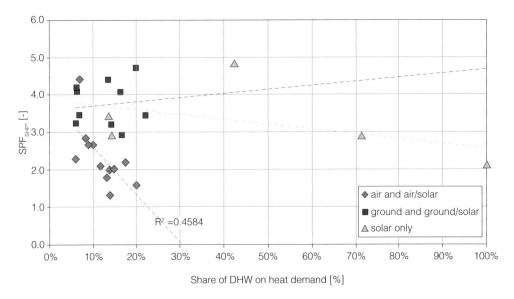

Fig. 6.6 Measured or correlated SPF$^*_{SHP}$ depending on the share of domestic hot water measured for solar and heat pumps with solar as only source

example, system control and installation quality. This becomes evident as some systems with a moderate collector area outperform even ground source installations.

Looking at well-performing systems of different configurations shows their potential. Parallel, series, and regenerative approaches are found among these convincing examples, and no significant difference could be seen on the main source for the heat pump system (Figure 6.8).

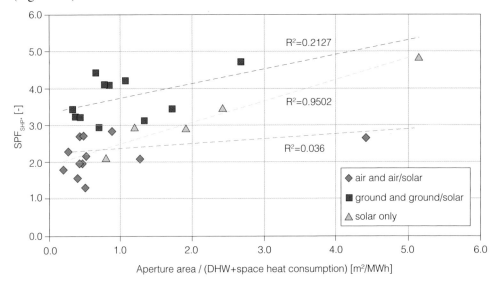

Fig. 6.7 Measured or correlated SPF$^*_{SHP}$ depending on the ratio of installed collector area and heat demand for solar and heat pumps with uncovered absorbers

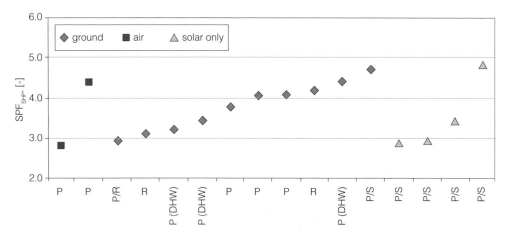

Fig. 6.8 Measured or correlated SPF^*_{SHP} of selected well-performing monitored systems with different concepts, grouped by the main sources air, ground, and solar only

6.4 Best practice examples

Besides gathering performance indicators, each of the monitored systems is a good source for knowledge creation. In the following, five systems were selected representing the variety of successful system solutions seen in the field:

- Blumberg, Germany: Parallel system with ground source heat pump and flat-plate solar collector (published in Ref. [3]).
- Jona, Switzerland: Parallel system with air source heat pump and flat-plate solar collector (published in Ref. [4]).
- Dreieich, Germany: System with ground source heat pump and recovery of the ground by PVT collectors (published in Ref. [5]).
- Savièse, Switzerland: Parallel/series system with ice storage and unglazed collectors as source (published in Ref. [6]).
- Satigny, Switzerland: Parallel/series system for multifamily building with uncovered collector as source (published in Ref. [7]).

The variety of the systems with good performances indicates that adapted, well-installed solutions for solar and heat pumps are available.

6.4.1 Blumberg

The system in the Blumberg single-family house is a typical example for a larger parallel system. A ground source heat pump plus solar thermal system has been installed in 2008. The system provides domestic hot water for four people and space heating energy distributed by a floor heating system. A 9.9 kW$_{th}$ ground source heat pump and 14.4 m^2 flat-plate collectors deliver heat to a 1000 l combi-storage covering space heating and domestic hot water. The heat pump unit contains also an electric heating element. The system has been monitored from July 2009 till June 2010. Electric and thermal energy flows were monitored for evaluating SPF_{bSt}.

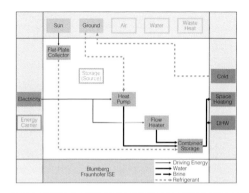
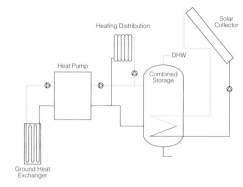

Location	Blumberg, Germany	Altitude 700 m
Building	Low energy, floor heating	276 m^2, 56 kWh/(m^2 year)
Heat pump	Ground source, scroll compressor, R410A	9.9 kW$_{th}$ COP = 4.5 @ B0/35, EN 255
Solar collector	Flat-plate collector	14.4 m^2, 20°E, inclination 40°
Thermal storage	Combi-storage	1000 l

For the period of 12 months, the solar collectors yielded 5005 kWh or 348 kWh/(m^2 year). The heat pump delivered 14 769 kWh to the storage, and the heating element 0.6 kWh. The total electricity demand of the system was 3437 kWh/year.

The seasonal performance factor amounted to SPF$_{bSt}$ = 5.8, shown in greater detail in Figure 6.9. The performance factor for the heat pump alone was SPF$_{HP}$ = 4.7. In the

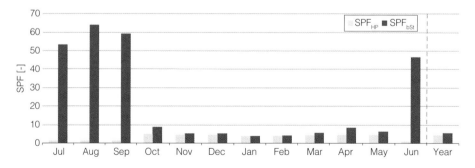

Fig. 6.9 SPF$_{bSt}$ on monthly and annual basis for the monitoring period of July 2009 till June 2010

6.4 Best practice examples

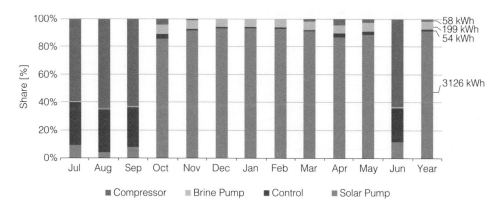

Fig. 6.10 Relative distribution of electricity consumption for SPF$_{bSt}$ boundaries on monthly basis. For annual results (rightmost column), absolute values are given as well

main summer month, the heat was delivered by solar collectors only. Their rather high performance shows the potential of the easy-to-control parallel system with a large collector subsystem.

The electrical energy consumption of the system (3437 kWh/year, seen with SPF$_{bSt}$ boundaries) can be divided into the consumption of compressor (3126 kWh/year), auxiliary electric heating (0.6 kWh/year), brine pump (199 kWh/year), control (54 kWh/year), and solar circuit pump (58 kWh/year). The relative distribution also on monthly basis is shown in Figure 6.10. The electrical heating element remained unused due to the reliable ground source.

Under the given circumstances such as the relatively cold but sunny climate, a low-temperature distribution, and a well-dimensioned solar system, a good performance could be achieved even with a more simplistic integration of the solar system.

6.4.2 Jona

The system in the single-family house located in Rapperswil-Jona, Switzerland, is a typical example for a larger parallel system. An air source heat pump plus solar thermal system has been installed in 2009 and monitored. The system provides domestic hot water (1400 kWh/year) for two people and space heating (18 700 kWh/year) for 200 m^2 of heated floor area of a house built in 1992. The 20 kW$_{th}$ air source heat pump and the 15 m^2 covered solar thermal collector field deliver heat to a tank-in-tank solar combi-storage of 1.8 m^3 water volume from where the needs for space heat and domestic hot water are served. From the solar combi-storage, domestic hot water and space heat are provided. The heat pump charges either the upper part or the middle part of the storage directly (switching with two three-way valves) and the solar thermal collector field charges the storage with internal heat exchangers placed in the top and in the bottom third, of which the top heat exchanger can be circumvent.

All heat inputs and outputs of the store were monitored from February 2010 to December 2011 as well as the electricity consumption of the heat pump and the solar collector operation. Electricity for all controllers and pumps with the exception of the space heat distribution pump was included in the balance. Measured points included the flow and

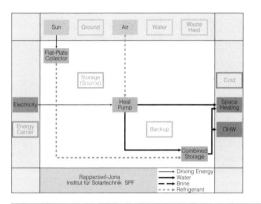
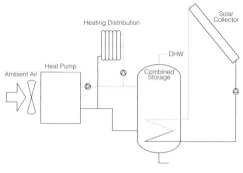

Location	Jona, Switzerland	Altitude 408 m
Building	Low energy, floor heating	$200\,m^2$, $100\,kWh/(m^2\,year)$
Heat pump	Air source, scroll compressor, R404A	$19.7\,kW_{th}$ COP = 3.8 @ A2/35, EN 255
Solar collector	Flat-plate collector	$15\,m^2$, 20°W, inclination 38°
Thermal storage	Combi-storage	1800 l

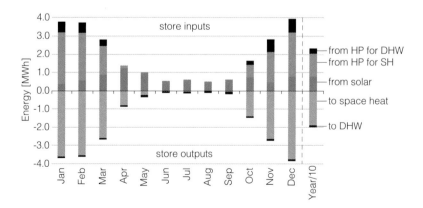

Fig. 6.11 Energy balance of the combi-storage

6.4 Best practice examples

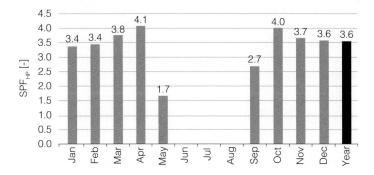

Fig. 6.12 Performance factors of the heat pump SPF$_{HP}$

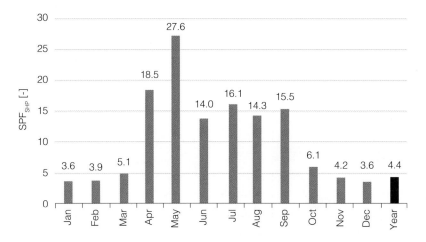

Fig. 6.13 Performance factors of the system including the storage SPF$_{bSt}$

return temperatures and volume counts of the solar circuit, the heat pump circuit and the space heat circuit, and cold water and hot water temperatures and tapped volumes. For the year 2011, the resulting seasonal performance factor of the system calculated based on all electricity use and the useful heat leaving the store was SPF$_{SHP}$ = 4.4. Compared with a typical air source heat pump, this is a strong improvement of the performance that mainly results from the high share of solar heat in the heat balance (Figure 6.11). The low temperatures in the heating circuit together with the low share of DHW lead to low temperatures at the condenser, so even in winter monthly COP for the air source heat pump of >3.4 was achieved (Figures 6.12 and 6.13).

6.4.3 Dreieich

A heat pump system with uncovered PVT collectors and ground heat exchanger has been measured over a period of 2 years starting in March 2009. It consists of a 39 m² PVT collector field, a 12 kW heat pump, and a total length of 225 m coaxial borehole HE. The system provides space heat and domestic hot water for a large single-family house. The measured heat demand in the system is 36 MWh/year in the first and

40 MWh/year in the second year of operation, which corresponds to 25 and 45% increased heat demand compared with planning. The additional PV yield due to PV cooling is determined by comparison with non-cooled PV reference modules. In both measured periods, the additional PV yield due to cooling is 4%. The thermal yield of the PVT collector is 450 kWh/(m² year).

The PVT collector connection to the heat source side aims at two effects compared with a conventional heat pump system. One effect is the lift of the cold temperature level by the thermal yield of the unglazed collector. Simultaneously, the PV is cooled and therefore its efficiency and electrical yield are improved. The collector is connected to the heat source by a simple switching valve. This valve allows connecting the PVT collector in series to the borehole heat exchanger. Accordingly, the supplied heat from the collector is restricted to the heat source side of the heat pump. In the monitored system, the heat pump provides heat for the space heating system and the DHW storage. The DHW storage is charged by the integrated condenser of the heat pump that can be operated as a desuperheater.

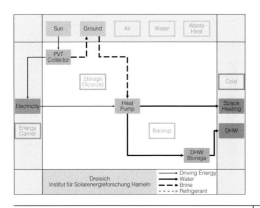
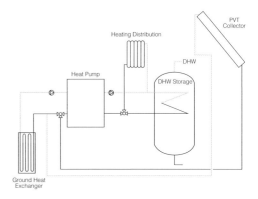

Location	Dreieich, Germany	Altitude 150 m
Building	Low energy, floor heating	380 m², 66 + 6 kWh/(m² year)
Heat pump	Ground source, scroll compressor, R407C	11.6 kW$_{th}$ COP = 4.8 @ B0/35, EN 255
Solar collector	PVT collector	39.6 m², 24°E, inclination 15°, 2.8 kW$_p$
Thermal storage	DHW	150 l

6.4 Best practice examples

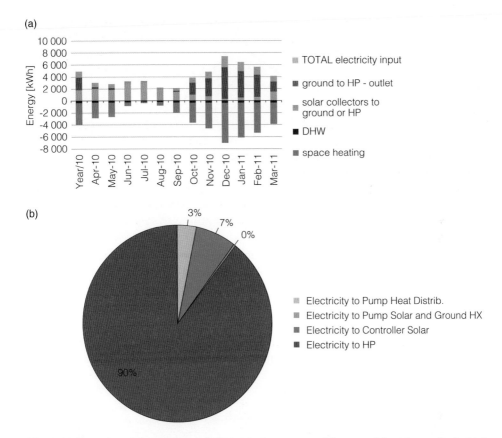

Fig. 6.14 Heat demand and energy balance in the course of the year (a) and overall electricity consumption (b) in the second year of operation. The monthly heat flow rates of the ground heat exchanger are monthly positive net heat flows. The solar collector yield, which exceeds the evaporator demand, is injected to the ground heat exchanger

The results of the measurements are given in Figure 6.14. It shows the seasonal distribution of the heat fed into the boreholes and the extraction of this heat by the heat pump in winter. The small share of DHW gives optimal low temperatures for the heat pump operation.

Figure 6.15 shows that performance and temperature level of the ground heat exchanger are comparatively unchanged, because of the active ground regeneration by the solar thermal PVT collector. The comparatively good stability of heat source temperatures highlights the robustness of the combined heat source unglazed solar collector and borehole heat exchanger.

The PVT collectors have been installed with and without rear side insulation. The measurement was conducted separately for both fields without significant differences for electrical and thermal yields.

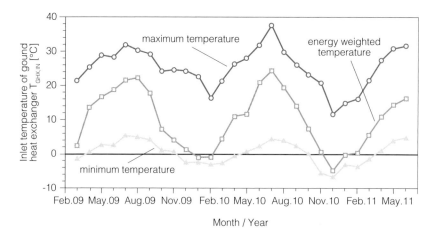

Fig. 6.15 Monthly maximum, minimum, and average temperatures of the ground heat exchanger inlet. The minimum temperatures in winter 2010/2011 are caused by nonstop operation of the heat pump over a period of 2 weeks

6.4.4 Savièse

The system installed in an old retrofitted house in Savièse-Granois in Switzerland is made for DHW production and space heating up to 8 kW. The selective coated unglazed

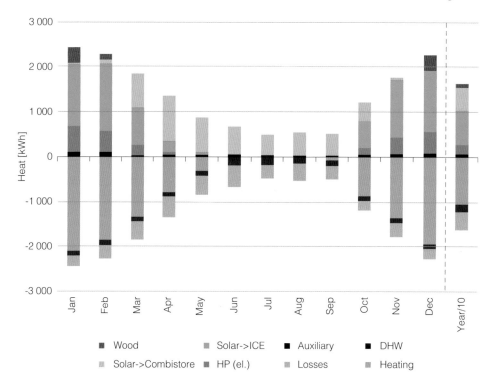

Fig. 6.16 Heat balance on monthly and annual basis

6.4 Best practice examples

collectors are used both as solar collector and as heat exchanger on ambient air. They charge a storage tank with phase changing material (PCM; ice/water) connected in series with the collector field on the brine/water heat pump evaporator. When running at higher temperatures, they charge a combi-storage tank that is connected to the heating distribution and produces DHW.

The results and the monthly energy balance are shown in Figure 6.16. With an annual SPF_{SHP} of 4.3 (3.3 when counting the additional heat from a wood stove as electricity), the installation can be compared with standard heat pump with boreholes. The concept of the ice storage in combination with the uncovered collectors is promising especially when ground sources are not available. In addition, the noise impact of an external air unit can be avoided.

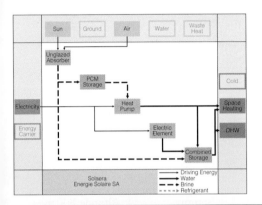
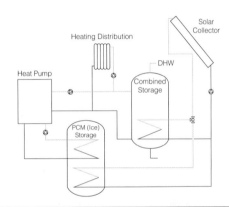

Location	Savièse, Switzerland	Altitude 1500 m
Building	Retrofitted low-energy building	230 m², 43 + 6 kWh/m²
Heat pump	Brine/water with "solar only" source, R407C	6.9 kW$_{th}$ COP = 4.3 @ B0/35, EN 255
Solar collector	Type "AS" from Energie Solaire SA (uncovered)	30 m², 10°E, inclination 20°
Thermal storage	PCM (ice/water) storage Combi-storage	620 l 1000 l

6.4.5 Satigny

The system described in this section is implemented in a new housing complex (9552 m²) in Satigny near Geneva, Switzerland. This housing complex is composed of 10 similar blocks of 8 flats each. Each block (approximately 950 m²) has its own heating system, designed to produce space heating (floor heating with flow temperatures around 30 °C) and domestic hot water. A detailed energy monitoring was implemented in one of the blocks, and enabled to describe the behavior of the system, for example, consumption, control strategy, and temperature levels, and to evaluate its performance over an entire year.

Each heating system consists of a heat pump directly coupled to 123 m² uncovered solar collectors as its heat source. As seen in the figure below, heat can be delivered to the flats:

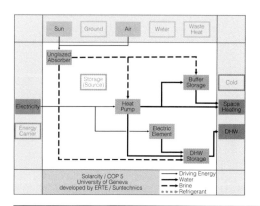
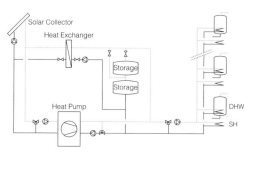

Location	Satigny, Switzerland	Altitude 500 m
Building (one block)	Minergie label Floor heating 32 habitants	927 m²
Heat pump	Scroll compressor with enhanced vapor injection R407C	35 kW$_{th}$ COP = 4.5 @ B0/35, EN 14511
Solar collector	Type "AS" from Energie Solaire SA (uncovered)	60 × 2.05 m², 37°E, 20° slope
Thermal storage	Water buffer DHW	6000 l per block 300 l per flat

6.4 Best practice examples

first directly from the solar collectors, second from the 6000 l storage (if temperature from the solar collectors is too low), third from the heat pump (if temperature in the storage is too low), and fourth by backup electric heating, placed in the storage, in case of a regular heat pump shutdown (outdoor temperature below $-20\,°C$) or failure.

In summer nights, when cooling is appreciated, heat from the flats can be dissipated via the solar collectors. Another particularity of this system is the hydraulic distribution circuit. The same circuit alternatively delivers space heating and DHW to the flats. Since space heating and DHW cannot be delivered simultaneously, each flat is equipped with a 300 l DHW buffer storage.

Operation of the system on a typical mid-season day is represented in Figure 6.17. During this specific day, the average outdoor temperature is $7\,°C$ and the maximum solar irradiance is $600\,W/m^2$. The demand is characterized by (a) four DHW cycles of about

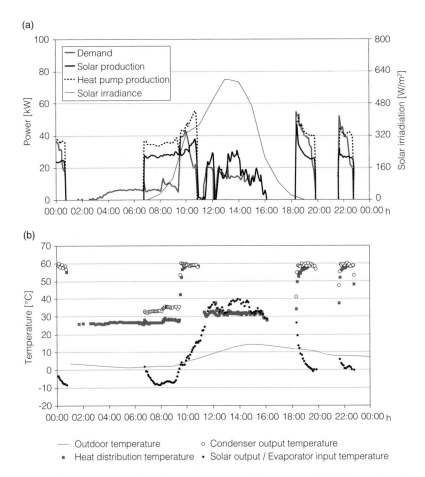

Fig. 6.17 Mid-season day (March 10, 2013, 5 min time step). (a) Demand, solar and heat pump power (primary axis), and global solar irradiance (secondary axis). (b) Outdoor, solar output, HP output, and distribution temperatures

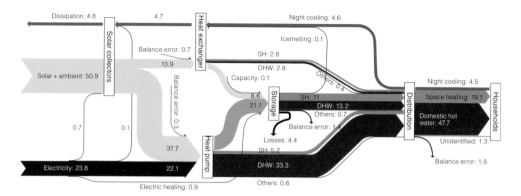

Fig. 6.18 Sankey diagram of the studied block for the year 2012 (values in kWh/(m² year))

1 h for charging of the individual DHW storages (starting at 0:00, 9:30, 18:00, and 21:30 h), during which heat is distributed at about 60 °C; and (b) two space heating cycles (2:00–9:30 and 11:30–16:00 h), during which heat is distributed between 25 and 30 °C.

The four DHW cycles are covered by the heat pump, with an absorber output/evaporator input temperature between −10 and 20 °C. The excess production, in particular at the end of each cycle, is stored in the central buffer tank. As a result of such storage, space heating starts with storage discharge (2:00–6:30 h), followed by heat pump production. In the afternoon, solar irradiance is high enough for space heating to be covered directly by the collectors (heat pump off), with a solar output temperature above 30 °C.

The detailed monitoring enabled to fully characterize the energy flows of the studied block over the entire year of 2012 (Figure 6.18), with an input–output balance error of less than 3%.

Thanks to the high-performance thermal envelope and to controlled internal temperatures (average of 20 °C), the space heating demand (19 kWh/(m² year)) is low compared with standard values in Switzerland. On the contrary, DHW consumption (48 kWh/(m² year)) is higher than usual, which is due to a high occupancy rate (as a matter of fact, the specific user consumption of 1184 kWh/inhabitant per year is only slightly above the average of 1075 kWh/inhabitant per year). All in all, this results in an unusual space heating and DHW ratio of 30:70, which is expected to become common in high-efficient multifamily dwellings.

The thermal storage plays an important role in the system, since 37% of the energy supplied to the flats goes through the storage before use (44% in summer). The storage heat losses amount to 14% of the storage input energy.

The renewable heat fraction is 68% of the total energy input, and the performance factor of the heat pump SPF_{HP} is 2.7. Note that the monthly SPF_{HP} only varies slightly during the year (2.5–3), even in summer. Indeed, during this period, the heat pump works only to produce DHW at 60 °C (when direct solar is not enough), that is, at high condenser temperature. As a result, the overall SPF_{SHP} only amounts to 2.9 (2.5 in winter and 4.4 in summer). As it relates to a low heat demand (57 kWh/(m² year)), the total electricity consumption, however, remains within an acceptable value of 24 kWh/m².

References

1. Miara, M., Günther, D., Langner, R., and Helmling, S. (2014) The outcomes and lessons learned from the wide-scope monitoring campaign of heat pumps in family dwellings in Germany. Proceedings of the 11th IEA Heat Pump Conference, May 12–16, Montréal, Canada.

2. Loose, A. and Drück, H. (2013) Field test of an advanced solar thermal and heat pump system with solar roof tile collectors and geothermal heat source. Proceedings of the 2nd International Conference on Solar Heating and Cooling for Buildings and Industry (SHC), September 23–25, Freiburg, Germany. *Energy Procedia*, **48**, 904–913.

3. Ruschenburg, J., Palzer, A., Günther, D., and Miara, M. (2012) Solare Wärmepumpensysteme in Einfamilienhäusern – Eine modellbasierte Analyse von Feldtestdaten. Proceedings of the 22nd Symposium "Thermische Solarenergie", May 9–11, Bad Staffelstein, Germany.

4. Haller, M. and Frank, E. (2012) System-Jahresarbeitszahl größer 4.0 mit Luft-Wasser Wärmepumpe kombiniert mit Solarwärme. Proceedings of 22nd Symposium "Thermische Solarenergie", May 9–11, Bad Staffelstein, Germany.

5. Bertram, E., Glembin, J., and Rockendorf, G. (2012) Unglazed PVT collectors as additional heat source in heat pump systems with borehole heat exchanger. Proceedings of the 1st International Conference on Solar Heating and Cooling for Buildings and Industry (SHC), July 9–11, San Francisco, CA, USA. *Energy Procedia*, **30**, 414–423.

6. Graf, O. and Thissen, B. (2012) Chauffage par pompe à chaleur solaire avec des capteurs sélectifs non vitrés et accumulateur à changement de phase. Final report, programme de recherche: SI/500'481 Swiss Federal Energy Office, Bern, Switzerland.

7. Fraga, C., Mermoud, F., Hollmuller, P., Pampaloni, E., and Lachal, B. (2012) Direct coupling solar and heat pump at large scale: experimental feedback from an existing plant. Proceedings of the 1st International Conference on Solar Heating and Cooling for Buildings and Industry (SHC), July 9–11, San Francisco, CA, USA. *Energy Procedia*, **30**, 590–600.

7 System simulations

Michel Y. Haller, Daniel Carbonell, Erik Bertram, Andreas Heinz, Chris Bales, and Fabian Ochs

> **Summary**
>
> Different concepts for solar and heat pump (SHP) heating systems were evaluated based on annual system simulations within Subtask C of T44A38. A direct comparison of simulations carried out with different climatic conditions or heat loads is not possible. The reason for this is that the influence of these "boundary conditions" on the performance of both heat pumps and solar collectors is very strong. Therefore, a set of common boundary conditions for the simulation of SHP systems in different European climates has been defined for T44A38. These boundary conditions include climate, heat sources, and useful heat to be delivered (Section 7.A), as well as the procedures for the determination of performance figures (Chapter 4). Only this shared set of boundary conditions allows reliable and reproducible comparison of system concepts simulated by different working groups. The international collaboration within T44A38 has led to 12 reports on simulations where the common boundary conditions have been applied on different simulation platforms, and 17 reports where other boundary conditions were applied. This large amount of simulation results enables to derive general conclusions for the investigated system concepts. As a result, it is concluded that parallel air source SHP systems can be as efficient as ground source heat pumps without solar collectors. Parallel ground source SHP systems achieve the highest performance factors. In many simulation studies, the seasonal performance factor (SPF) of the system increased by 1–2 by using solar heat for domestic hot water and space heating. Series SHP systems achieve similar performance to parallel SHP systems with air source heat pumps or ground source heat pumps without solar collectors. Systems with only solar collectors or absorbers as a heat source were found to be able to perform in the range of air source SHP systems. Ground regeneration for single boreholes was not found to increase system performance significantly if the boreholes were not undersized. For heat pumps in combination with combi-storages, it is shown that storage stratification efficiency and proper hydraulic integration and control have a large influence on the energetic performance.

7.1 Parallel solar and heat pump systems

The simplest, and so far the most frequently used, solutions for combining solar thermal and heat pump systems are purely parallel concepts. In these systems, both solar collectors and the heat pump deliver heat at a temperature level that can be used directly, or after storage at a temperature that is suitable for direct use. Examples for the possible energy flows in these concepts are shown in Figure 7.1 for a system where solar heat is

Solar and Heat Pump Systems for Residential Buildings, First Edition.
Edited by Jean-Christophe Hadorn.
© 2015 Ernst & Sohn GmbH & Co. KG. Published 2015 by Ernst & Sohn GmbH & Co. KG.

160   7 System simulations

(a)

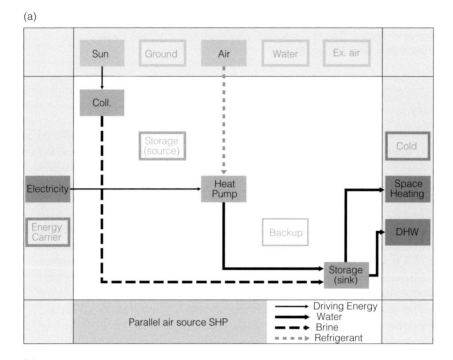

(b)

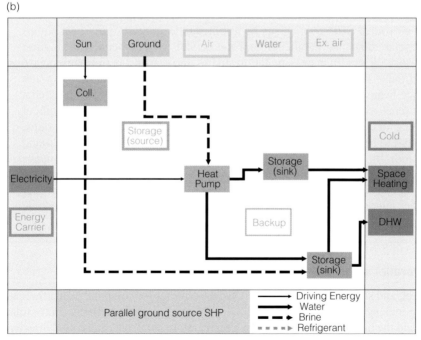

Fig. 7.1 Energy flow chart with parallel air source (a) and ground source (b) SHP systems for solar DHW and space heating (a) and for solar DHW only (b)

7.1 Parallel solar and heat pump systems

used for domestic hot water (DHW) and space heating (a) and a system where solar heat is used only for DHW (b).

It is tempting to assume that such a parallel concept is simple and not much different from the combination of solar thermal collectors with oil or gas boilers. However, this is not the case! A vast amount of data collected from field and laboratory measurements as well as from simulations within T44A38 shows that there are fundamental differences between these systems. In particular, the design of parallel systems needs to consider the sensitivity of the heat pump to increased temperatures on the sink side as well as the high mass flow rates that pumps are operated with, which may be unusual for solar storage tanks (see Section 3.5).

7.1.1 Best practice for parallel solar and heat pump system concepts

Recommendations given for the integration and control of heat pumps into hydronic heating systems that are given, for example, in Ref. [1] also apply for combined solar and heat pump systems. In particular, the dependence of the coefficient of performance (COP) of heat pumps on the temperature of the heat sink[1] requires a careful planning and engineering of all components in order to avoid mixing processes.[2]

In general, storing heat in a water storage increases the temperature level that needs to be delivered by a heat pump or solar collectors because of several reasons:

- heat exchangers for transferring the heat into or out of the storage (if needed);
- the nature of sensible storage (storing by increasing the temperature);
- heat losses of the storage;
- mixing effects in the storage.

Heat storage should therefore only be used if there is a substantial advantage from storing heat. In general, a direct supply of heat from the heat pump to the space heat distribution is desirable. At the same time, frequent on/off cycling of the heat pump compressor must be avoided in order to achieve high lifetime of these units. For this reason, a heat storage is usually used in the space heating loop in order to decrease the on/off cycling of heat pumps. A second motivation for thermal storages is the necessity to bridge the heat supply for pauses of electric supply, which are part of widespread and economically attractive electric heat pump tariffs.

For low-temperature space heat distribution systems that have a high thermal capacity (e.g., floor heating systems), it is sometimes recommended to refrain from the general practice of using thermostatic valves in combination with heat pumps. Thus, the on/off cycling frequency of the heat pump may be low enough even without a technical storage in the space heating loop. However, thermostatic valves control the heat input into the rooms, and thus they reduce space heating demand effectively. Refraining from using thermostatic valves in SHP systems may lead to a higher seasonal performance factor (Figure 7.2). However, the electricity demand increases. The reason for this increasing electricity demand despite an increasing SPF for systems without thermostatic valves is

---

[1] See Section 3.2 on heat pumps.
[2] See also Section 3.5.2 on exergetic efficiency and storage stratification.

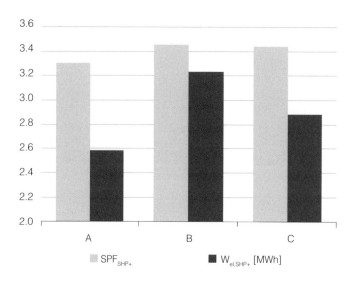

Fig. 7.2 Performance of a solar combi-system with different space heat distribution control: (A) with thermostatic valves (TVs), (B) without TVs, and (C) without TVs and heating curve reduced to the lowest possible values that achieve the minimum room temperature required; based on TRNSYS simulations with a one-zone building model. B and C use more electricity due to more heat delivered for space heating

that the amount of heat supplied to the heat distribution increases more than the electricity demand.[3] It has to be noted that a higher SPF is not an appropriate target function for system optimization in this case.

Two typical parallel SHP concepts that use solar heat for DHW and a heat pump as a DHW backup are as follows:

- A DHW storage tank that has an upper heat exchanger for charging by the heat pump and a lower heat exchanger for solar heat input. Care has to be taken that the heat transfer capacity of the upper heat exchanger matches the heating power of the heat pump. *The general recommendation is $0.4\,m^2$ heat exchanger surface per kW of the heat pump in summer (i.e., at A20W50), and that charging by the heat pump does not raise the temperature in the lower part that is reserved for solar heating.*
- A buffer storage and an external DHW module. Attention has to be paid to the control of the DHW module in order to always achieve low return temperatures to the storage. In this case, the heat pump charges the buffer storage directly.

In both cases, solar heat can be charged with an internal heat exchanger, mantle tank design, or an external heat exchanger.

---

[3] In other words, field measurements that show higher SPF for systems without thermostatic valves are not a valid proof for absolute electricity savings and a better energetic performance of these systems in comparison with systems with thermostatic valves.

7.1 Parallel solar and heat pump systems

Systems that use solar heat for both DHW and space heating use either two separate storage tanks – one for DHW and one for space heat – or one combi-storage. The first solution has the advantage of a clear separation between the different temperature levels of space heat and heat for DHW. Combi-storages, on the other hand, have the advantage that there is only one storage to which heat has to be transferred from the solar loop.[4] This reduces the complexity of the solar integration and the number of components that have to be installed, and provides a better surface to volume ratio, which reduces heat losses. However, for low-temperature space heating systems the stratification efficiency on the one hand and the integration and control of the heat pump into systems with solar combi-storages on the other hand are crucial for the efficiency of these systems. Excessive mixing in the storage unit as well as a poor integration and control of the heat pump may lead to an increase of the annual electricity demand on the order of 50% [2]. The main reason for this is that heat prepared by the heat pump in DHW mode (charging the DHW zone of the combi-storage) is subsequently mixed with fluid at lower temperatures and then used for space heating instead of DHW. If this is the case, the heat pump is forced to deliver an excessive amount of heat in DHW mode with a significantly lower COP than in space heating mode.

Recommendations for the combination of heat pumps with solar combi-storage (based on Refs [2,3]; see also Figure 7.3)

1. The position of the DHW sensor for boiler charging control must be placed at a safe distance from the space heating zone of the storage:
   a. This distance is system-specific (it depends on the stratification capabilities of the storage).
   b. As a first guess, a minimum distance of 30 cm is recommended.
2. The return line from the storage to the heat pump in DHW mode must be placed above the space heating zone of the storage.
3. Direct storage inlets have to be designed for high flow rate charging without disturbing the temperature stratification within the storage. The maximum flow rates that do not disturb stratification should be declared by the manufacturer.[5]
4. Charging the DHW zone by the heat pump should be restricted to one (maximum two) time window(s) of not more than 2 h/day. For a standard tapping profile, the best time for this window is between 4 p.m. and 8 p.m. [4], or for economic optimization within a time period of lower electricity price.
5. It can be advantageous to bypass the storage when the heat pump runs in the space heating mode.

---

[4] See also Figure 3.7 on different variants for DHW preparation from combi-storages.
[5] See also Figure 3.11 on CFD simulation results of a combi-storage.

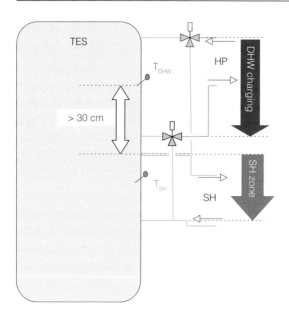

Fig. 7.3 Recommended hydraulic solution for the combination of heat pump and combi-storage (solar and DHW integration are not shown). Ref (Haller et al.) Haller, M.Y., Haberl, R., Carbonell, D., Philippen, D. & Frank, E., 2014. SOL-HEAP - Solar and Heat Pump Combisystems. Report Contract number SI/500494-02, Institut für Solartechnik SPF, Hochschule für Technik HSR, Rapperswil, Switzerland.

7.1.2 Performance of parallel solar and heat pump systems

The seasonal performance factor of the overall system (SPF_{SHP+}) increases significantly if solar thermal heat is added in parallel to a heat pump for providing DHW or for providing DHW and space heat. The reason for this is that part of the heating demand is covered by the solar collectors that have a much higher ratio of heat delivered to electricity consumed (COP_{SC} around 40–300) than the heat pump (COP_{HP} around 2–6).

The increase in SPF_{SHP} is largely dependent on the heat load (total heat demand, share of DHW, and distribution over the year) as well as on the solar resource that is available (climate and collector area and orientation). For a typical single-family house in a Central European climate and low solar fractions, an increase in SPF_{SHP+} in the range of 0.1 SPF point per m^2 collector area can be expected. Most relevant from an investor's perspective are the resulting overall electricity savings of the system. Although the thermal collector yield is often in the range of 400 (combi-system) to 600 kWh/m^2 (DHW), the achieved electricity savings are much lower because the heat pump delivers heat with an SPF in the range of 2.5–5. This is illustrated in Figure 7.4, where the specific collector yield is in the range of 250–600 kWh/(m^2 year), but the electricity savings are only in the range of 60–140 kWh$_{el}$/m^2, that is, roughly four times lower.

7.1 Parallel solar and heat pump systems

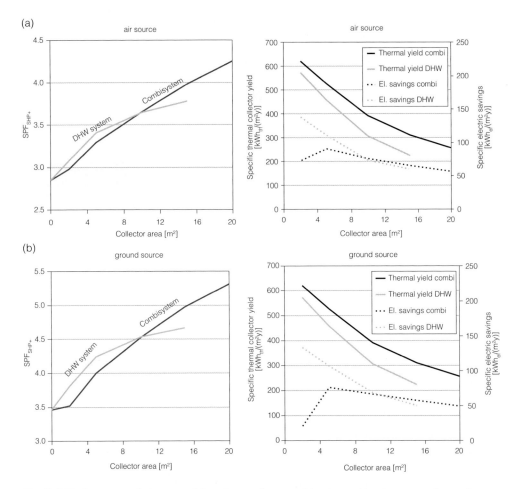

Fig. 7.4 Performance of air source (a) and ground source (b) solar and heat pump systems where solar system is used for DHW only (DHW) or for a combi-system with a combi-storage (combi). Specific yields and savings are per m² collector area

A general rule for thermal systems in combination with fuel burning devices is that collector yield and consequently also fuel savings decrease with increasing temperatures of the heat demand. In contrast, in heat pump systems, increasing temperatures of heat demand reduce the COP of the heat pump more than they affect the collector efficiency, and thus the electricity savings compared with a "heat pump only" system may well increase with increasing temperatures of the heat demand.[6]

---

[6] A decrease of the COP by 2–3%/K may be assumed for the heat pump, whereas <1%/K may be assumed for the efficiency of a flat-plate collector at an irradiation level of 800 W/m² (1.5%/K @ 400 W/m²).

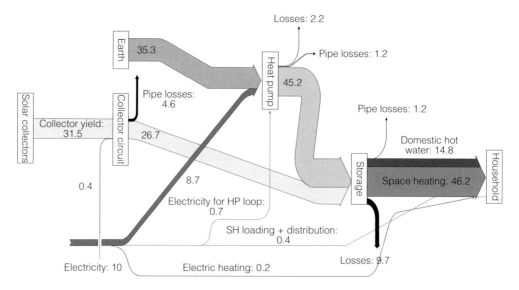

Fig. 7.5 Sankey diagram for a ground source SHP system with 15 m² collector area for the heat load SFH45 in Strasbourg. All values are given in kWh/year and per m² heated floor surface

In SHP systems with air source heat pumps, the SPF of the heat pump itself usually decreases slightly with increasing collector area of a solar combi-system. In this case, solar heat replaces heat pump space heating predominantly in transition periods when the COP of the heat pump is high (higher ambient temperature and lower space heat temperature requirements when the sun is shining). For solar DHW systems, the opposite is true. This is a result of less DHW preparation by the heat pump. Consequently, the SPF of the heat pump, which is a combination of the SPF in DHW mode and in space heating mode, is increasingly dominated by the higher COP of the space heating operation.

Thus, although the thermal collector yield is usually higher for solar combi-systems than for solar DHW systems (for a given collector area), the electricity savings for combi-systems remain lower until a collector area is reached that is oversized in comparison with the DHW demand (in Figure 7.4 around 10 m²).

The energy flows of a typical parallel SHP system with a ground source heat pump and a combi-storage are shown in the Sankey diagram of Figure 7.5.

7.1.3 Performance in different climates and heat loads

Climates and heat loads have a large effect on the performance of solar and heat pump systems. Figure 7.6 shows seasonal performance factors of the system (SPF$_{SHP}$) as well as electricity savings for solar combi-systems compared with a reference system without

7.1 Parallel solar and heat pump systems

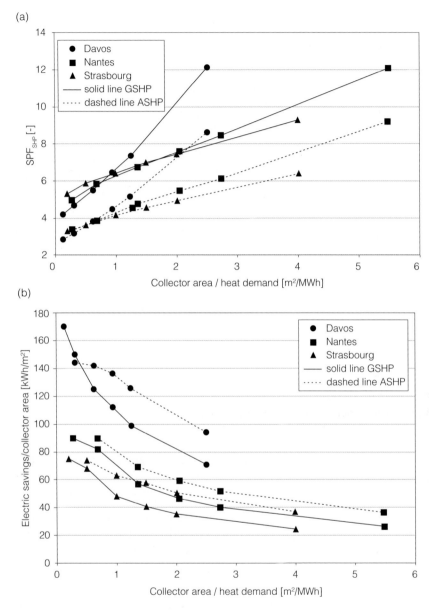

Fig. 7.6 Seasonal performance factor of the system (SPF$_{SHP}$) and electricity savings per m^2 collector area for ground source and air source solar and heat pump systems in different climates and with different collector areas of 2, 5, 10, 15, 20, and 40 m^2 (recalculated from Ref. [5])

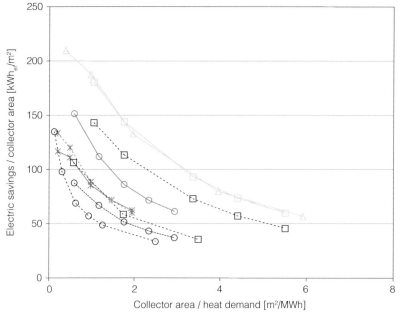

- △ DHW ASHP Innsbruck MFH (Haller)
- □ DHW ASHP Innsbruck MFH (Ochs)
- □ DHW WSHP Innsbruck MFH (Ochs)
- ○ combi ASHP Strasbourg SFH45 (Carbonell)
- ✻ combi ASHP Strasbourg SFH 60 (Haller)
- ○ combi GSHP Helsinki SFH45 (Carbonell)
- □ combi GSHP Strasbourg SFH45 (Bertram)
- ○ combi ASHP Strasbourg SFH45 (Carbonell)
- ✻ combi ASHP Strasbourg SFH 60 (Haller)

Fig. 7.7 Electricity savings achieved per m² collector area for different heat demand specific collector areas

solar collectors for different climates and for different collector areas.[7] The performance figures are plotted against a heat demand specific collector area (i.e., m² collector area per MWh heat demand). The total heat loads ranged from 7300 kWh/year for Nantes to 16 000 kWh/year for Davos, and the collector areas were from 2 to 40 m².

Depending on climate and collector area, SPF_{SHP} reached values from 4 to 12 for ground source systems and from 3 to 8 for air source systems. Although the relative increase in SPF_{SHP} is higher for ground source systems than for air source systems, the absolute electricity savings are typically higher for air source systems. This is due to the fact that the electricity consumption of the reference systems without solar collectors is significantly higher for the air source than for the ground source heat pump.

Figure 7.7 shows results from different authors obtained with different simulation tools, both for solar DHW systems (light) and for solar combi-systems (dark). The figure

---

[7] Results based on Polysun simulations of air source and ground source systems by Carbonell *et al.* [5].

shows that quite large electricity savings are possible (>180 kWh/m²) for small DHW systems in multifamily houses.

7.1.4 Fractional energy savings and performance estimation with the FSC method

As shown in the previous chapter, the influence of the particular climate and heat load on the electric energy savings that can be achieved with a certain collector area is large. In the IEA SHC Task 26, the FSC[8] method was developed in order to estimate for a specific type or design of a solar heating system the fractional energy savings f_{sav} based on monthly data of solar irradiation and heat load [6,7]. The idea behind this concept is that for a given solar thermal system design, f_{sav} depends on the usable irradiation that is available on the collector field $Q_{solar,usable}$, divided by a reference energy demand. For SHP systems, we define this reference energy demand based on the useful heat delivered $Q_{tot} = Q_{SH} + Q_{DHW}$:[9]

$$FSC = \frac{Q_{solar,usable}}{Q_{tot}}. \tag{7.1}$$

The usable solar energy $Q_{solar,usable}$ is defined assuming that for each month of the year the usable solar energy cannot be larger than the energy demand of the same month:

$$Q_{solar,usable} = \sum_{i=1}^{12} \min(Q_{tot,i}, Q_{solar,i}), \tag{7.2}$$

where Q_{tot} is total heat demand over the whole year (kWh), $Q_{tot,i}$ is total heat demand for month i (kWh), and $Q_{solar,i}$ is total solar irradiation on the collector plane for month i (kWh).

Figure 7.8 shows fractional electricity savings for data from Ref. [5] that were presented in the previous chapter, for various climates and heat loads. The collector areas ranged from 2 to 40 m². FSC correlations are obtained by a polynomial fit of all data pairs where FSC < 1.0:

$$f_{sav} = a \cdot FSC^2 + b \cdot FSC. \tag{7.3}$$

An analysis of the simulation results from Ref. [5] showed that removing data from simulations of passive house concepts (SFH15) and low energy demand in general improved[10] the correlations considerably.

---

[8] Fractional solar consumption, that is, the fraction of utilizable solar resource divided by the total energy demand.

[9] In Refs [6,7], the total energy demand was taken as the fuel consumption of a reference system without solar collectors. However, for SHP systems, taking the final electricity demand of a heat pump only system is not compatible with the idea behind the concept, and therefore the final heat demand for DHW and space heat, that is, the useful heat delivered, was taken instead.

[10] R^2 increases from 0.93 to 0.97 for ground source and from 0.97 to 0.98 for air source systems.

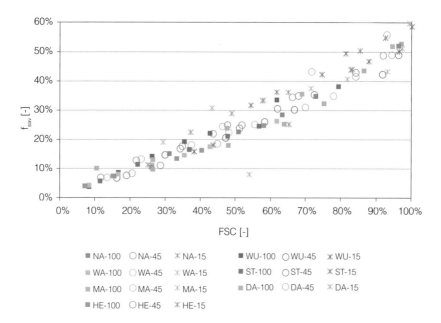

Fig. 7.8 Fractional electricity savings versus FSC for different climates and different heat loads for ground source SHP systems

Figure 7.9 shows the resulting correlations for a particular air source and for a particular ground source SHP system concept. It must be noted that the correlations that are obtained are still quite lower than those for the solar combi-systems with natural gas burner backup where $R^2 = 0.99$ has been reported in Ref. [7]. One reason for this may be that the performance of the heat pump depends not only on the monthly heat demand, but also on the temperature of the heat distribution, on the share of DHW in the heat load, and on ambient temperatures. Thus, the heat pump is influenced much more than a fuel boiler by

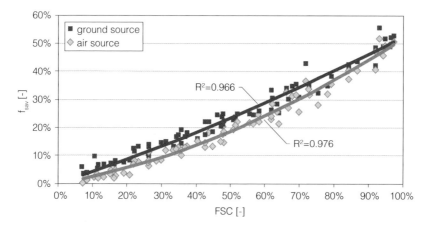

Fig. 7.9 Fractional electricity savings versus FSC for different climates and different heat loads

characteristics of the heat load that go beyond the total amount of heat needed per month. It also has to be kept in mind that the data shown in Figure 7.9 were obtained for one system concept, and results may look quite different for a differently designed SHP system.

7.2 Series and dual-source concepts

Series concepts use collector heat for the evaporator of the heat pump. In dual-source concepts, the heat pump may also use other sources than heat from the solar thermal collectors (e.g., air or ground). Single-source series concepts use exclusively solar collectors or absorbers as a heat source. The "single source" thereby does not refer to the solar irradiation, but to the collectors or absorbers. Thus, a system that uses only solar absorbers as a heat source is classified as a single-source system, even if the absorbers use ambient air as a heat source in addition to the solar irradiation. The intention of these concepts is to raise the temperature level of the heat source, or to downsize or completely replace an alternative heat source. For illustration of these concepts, the reader is referred to Figures 7.12–7.14.

7.2.1 Potential for parallel/series concepts with dual-source heat pump

In parallel/series (P/S) systems with dual-source heat pumps, collector heat can be used in parallel mode or in series mode, and the heat pump can use collector heat or an alternative heat source (dual source). Series collector operation may be either additional – that is, not reducing the amount and temperature of heat that is available in parallel operation mode – or substituting parallel collector operation at times when parallel operation would only be possible with low efficiency.[11]

Figure 7.10 shows results of an estimation of the theoretically available solar irradiation for additional series collector operation in a dual-source P/S system. Based on a climate data set for Strasbourg with hourly values for the global irradiation on a 30° tilted south oriented surface, the amount of solar heat that could be used in additional series operation is calculated using the following assumptions:

- collector efficiency data for a typical glazed flat-plate collector with selective coating;
- collector inlet temperature of 30 °C;
- collector mass flow rate of 15 kg/(m² h);
- direct use of solar heat is assumed, when the temperature lift in the collector reaches $\Delta T > 4$ K considering the above assumptions;
- series collector heat use is assumed if parallel use is not possible and if parallel use is not covering the whole load, that is, for months of September to April.

Monthly results are shown in Figure 7.10a. With the used assumptions, the available solar irradiation for series use is about 9% of the total solar irradiation onto the collector surface. Figure 7.10b shows the percentage of the total irradiation that is available for series use for four different climates and different collector inlet temperatures. With rising collector inlet temperatures in parallel mode, the collector efficiency decreases and thus also the irradiation that can be used in parallel mode decreases. Consequently, the potential for additional series use increases.

---

[11] See also Section 3.5.1.

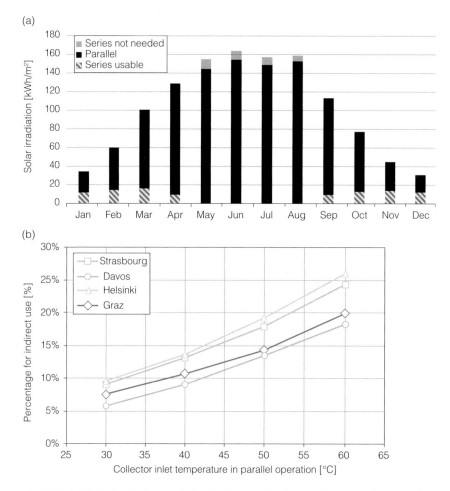

Fig. 7.10 (a) Solar irradiation available for direct and indirect collector use for a 30° tilted surface in Strasbourg and collector inlet temperature of 30 °C. (b) Percentage of the total irradiation available for indirect use for a 30° tilted surface for four different climates and different collector inlet temperatures

The potential for additional series collector heat use and for substituting parallel use with series use has been analyzed in Ref. [8] based on annual system simulations for different climates. Figure 7.11 shows limits for using heat from solar collectors for the evaporator of the heat pump instead of using an alternative heat source, in percentages of the total heat needed for the evaporator. In this analysis, solar heat was only counted as usable for the evaporator of the heat pump if

a) it was not possible to use the solar heat directly, but there was a demand for the evaporator of the heat pump on the same day and using the solar heat when it was available would have increased the COP of the heat pump compared with using the alternative heat source; these values are shown as "additional runtime potential";

7.2 Series and dual-source concepts

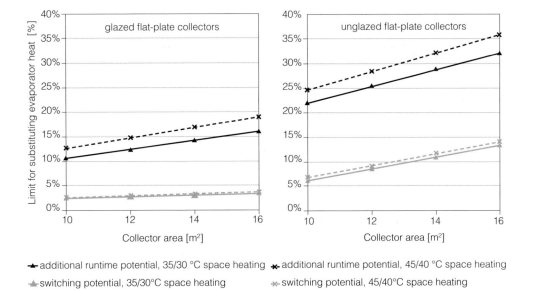

→ additional runtime potential, 35/30 °C space heating   ✶ additional runtime potential, 45/40 °C space heating
▲ switching potential, 35/30°C space heating   ✶ switching potential, 45/40°C space heating

Fig. 7.11 Theoretical limit of using heat from the solar collectors for the evaporator of the heat pump with benefit for the system's performance factor, given in percentage of the heat needed for the evaporator of the heat pump, for the climate of Zurich. (*Source*: Ref. [8].)

b) solar heat would normally have been used directly, but using it for the evaporator of the heat pump instead would have increased the overall COP of the system at this point in time; these values are shown as "switching potential."

For the type of system that was investigated, the additional series runtime potential was always higher than the potential for using solar heat for the evaporator at times when it would have been possible to achieve positive collector efficiencies also in parallel operation mode.

7.2.2 Concepts with Ground Regeneration

With boreholes of 80–300 m depth, the tapped heat reservoir is eventually regenerated from solar irradiation on the earth's surface and from the geothermal heat flux from the inner earth [9]. Continuous heat extraction from vertical boreholes may thermally deplete the ground in the time frame of decades. For single boreholes, the decrease in temperature level over time is small (around 1 K), but for larger borehole fields the effect superimposes and may be dramatic, if the boreholes are not sufficiently sized or actively recharged. With increasing time and depletion of the ground, the fraction of heat that is regenerated from the top rather than from the inner earth becomes more important and dominant. The effect of ground depletion is not stopped by estate boundaries and can therefore apply to neighboring ground heat exchangers in residential areas. Using solar heat in parallel to a ground source heat pump reduces the heat that is extracted from the ground and thus reduces indirectly the thermal depletion of the ground over decades. In addition, excess heat from solar collectors in summer or low-temperature heat from covered or unglazed collectors in general can be used to regenerate the ground actively. Different concepts have been presented for this purpose where either unglazed collectors or covered collectors are used (Figure 7.12).

The preferred way of studying effects of ground depletion and regeneration is by means of simulation models that are calibrated against measurements rather than by measurements only.[12] Based on the available literature on simulations of ground regenerating systems, the following recommendations can be given:

- Regeneration of single boreholes has little effect on the long-term temperature level of the heat source if the boreholes are dimensioned appropriately (not too short). In this case, the installation of thermal collectors or absorbers for the exclusive purpose of single borehole regeneration is unlikely to be economically attractive, and may use more pumping power electricity than can be spared by increase of the COP of the heat pump [10].
- For undersized[13] boreholes, regeneration may help to improve the system SPF [11,12].
- Large borehole fields may be depleted over the years by continuous heat extraction without regeneration, down to the point where they cannot be used as a heat source anymore. In this case, regeneration is essential.
- Heat from both covered and unglazed collectors should always be used first directly, and only excess heat (in summer) or low exergy heat (when the temperature for direct heat use cannot be reached in winter) should be used for regeneration. There is some evidence that this rule even applies for photovoltaic–thermal (PVT) collectors.
- The control strategy for switching from ground regeneration to direct collector heat use must be carefully chosen in order not to decrease the amount of direct heat use by the regeneration mode. Once the collector is operated in regeneration mode, the temperature of the collector will remain low, even though it could reach a temperature that is high enough for direct heat use if it would not be cooled by the ground heat sink.
- Only high-efficiency collector pumps are suitable for ground regenerating systems. Pressure losses for the operation at the low temperature levels must be evaluated carefully. This should also be kept in mind reading older reports and papers, which took into account far less efficient pumps than available today.

Further aspects are important for the design of these systems:

- High temperatures entering the polymeric ground heat exchanger can lead to serious damage and must be avoided.
- The collector field frost protection is usually designed for much lower temperatures than for the ground heat source loop. In a conjoint fluid loop, all limits must be respected, and it may be more appropriate to use two separate fluid loops with different frost protection (or no frost protection in the ground heat exchanger if possible).

---

[12] Measurements are not ideal in this case because of lacking reproducibility and the time frames involved.
[13] Undersized in comparison with the annual heat load.

7.2 Series and dual-source concepts

(a)
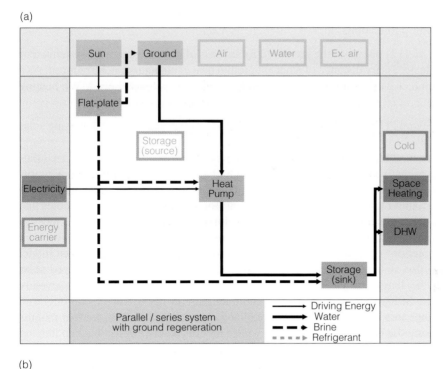

(b)
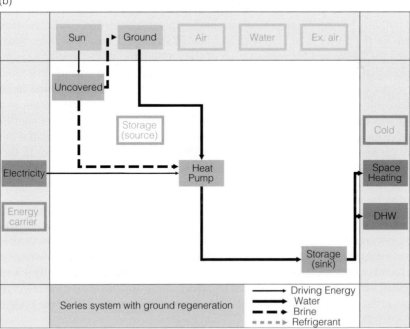

Fig. 7.12 Energy flow charts for a parallel/series/regenerative dual-source system (a) and for a series/regenerative dual-source system (b)

In addition to these general remarks, more detailed information is worth presenting from the following publications.

Bertram et al. [12] point out the increased robustness of solar regenerated systems. For instance, the necessary distance of adjacent ground heat exchangers can be reduced. As a rule of thumb, 1.2 m² of unglazed metal roof absorbers is proposed per MWh heating demand.

Kjellsson et al. [11] concluded for Swedish single-family home systems that using solar heat exclusively in parallel for DHW in summer and exclusively for partial ground regeneration in winter is the most effective solution to improve efficiency, if glazed flat-plate collectors are applied. In this case, in winter even higher evaporator temperatures are achieved than in pure series systems with regeneration. However, the advantage of complete ground recharging was only significant for small depths of the boreholes that required a direct electric heating backup during the coldest season of the year. Kjellsson et al, Bertram [10] simulated different SFH systems with a ground source heat pump and a single borehole of 110 m depth and an improved ground heat exchanger model respecting also short-term inertia effects. He concluded that by using unglazed solar collectors the length of the boreholes can be reduced by ~20% without severely affecting the SPF_{SHP+}. Reducing the borehole lengths further by increasing unglazed solar collector area had the side effect of reducing electricity demand in warmer months and increasing the demand in the coldest winter months where electric backup heating was increasingly used.

Ochs et al.[14] simulated different ground heat exchanger types (vertical, horizontal, basket, and building basement integrated/surrounding pipes) coupled to a solar and heat pump system for the T44A38 SFH15 with MATLAB/Simulink Carnot Blockset. Solar heat was used for DHW and for regeneration of the ground during the heating season. Care was taken that the use for ground regeneration did not reduce the solar contribution for DHW preparation. Electric backup was used for DHW preparation. Quite large collector fields of 10–30 m² were used, considering that only DHW preparation and regeneration was the intended use. It was concluded that regeneration did not significantly improve the performance of the systems, with the exception of the case of a flat ground heat exchanger integrated into the basement of the building structure. In this case, solar regeneration helped to reduce the additional space heating that was required because of the increased heat losses through the basement of the building caused by the heat extraction by the heat pump.

The effect of condensation gains on the annual unglazed absorber yield for ground regenerating systems has been analyzed in Ref. [13] for Central Europe based on models calibrated with data from field monitoring. Although for some winter days condensation gains were found to be responsible for as much as 30% of the collector yield, only 4% of the annual collector yield could be attributed to condensation and neglecting the effect of condensation in the annual simulations had no significant effect on the SPF of the system (difference < 0.02).

---

[14] Publication accepted for publication for the Proceedings of the Heat Pump Conference 2014 in Montreal.

An interesting concept for a ground regenerating system is currently investigated in Switzerland with boreholes that reach 300–500 m deep, of which the first 100–150 m are insulated [14]. It is claimed that this kind of borehole requires a new coaxial type of geothermal heat exchanger with increased thermal resistance between the downward and the upward flowing fluid in order to provide source temperatures in the range of 18 °C. The borehole is regenerated with heat from PVT collectors. In combination with these developments, a new low temperature lift heat pump is currently developed with the goal to reach a COP of >10 when producing space heat at 30 °C with the mentioned source temperature [15].

7.2.3 Other series concepts: dual or single source

Many of the series concepts include a thermal storage on the cold side to compensate the fluctuating supply of solar radiation. This storage may be an ice storage, where the phase change process from water to ice is used to provide heat for the heat pump, and the storage is regenerated, that is, the ice is melted, by solar heat. Also, storage tanks that are filled with an antifreeze mixture of water and glycol (glycol storage) may be used as cold side storage. Glycol storages have the advantage that no heat exchangers are needed since glycol is already used in the collector loop and on the source side of the heat pump. On the other hand, the glycol itself is an additional cost compared with water, and the storage capacity in the range of 0 °C is far lower than the phase change enthalpy of water/ice and thus a higher storage volume is needed for a glycol storage to store the same amount of heat. Energy flow charts of typical series systems are presented in Figure 7.13.

A first systematic analysis on the performance of different concepts for (covered) solar and heat pump systems for space and DHW heating was presented by Freeman *et al.* [16]. In this study, the three concepts of parallel, P/S dual source, and P/S single source were simulated with TRNSYS for three different climates in the United States and it was concluded that the thermal performance of the parallel concept was consistently superior to the other two concepts. This is surprising from a theoretical point of view, because an ideal controller in a dual-source system should switch to series operation, only if higher performance is achieved. Accordingly, P/S dual-source systems should perform at least equally to parallel systems, possibly by not operating the series connection at all. More than 30 years have passed since. During this time period, heat pumps and solar thermal collectors have become standard components for residential heating applications in Europe, and their performance is today far better than in the 1970s.

Citherlet *et al.* [17] studied different series and parallel/series concepts for three different Swiss climates. In this study, the best performing systems from an energetic point of view were systems with glazed flat-plate collectors that used the heat predominantly parallel to the heat pump. Whereas adding solar thermal collectors parallel to the heat pump improved SPF_{SHP} by 0.5, additional series heat use improved the SPF further by 0.1. When comparing two installations with the same collector area, a series/parallel system with the cold side storage and another system only for domestic hot water in parallel mode, the first system gives better performance for a collector area of $\geq 30\,m^2$ only. Using unglazed selective absorbers instead of glazed flat-plate collectors in these

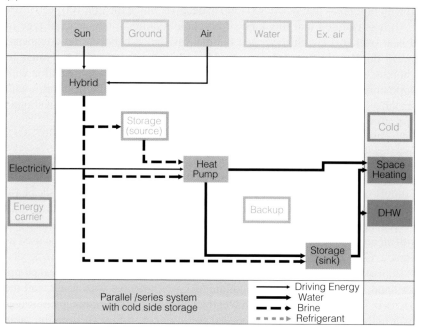

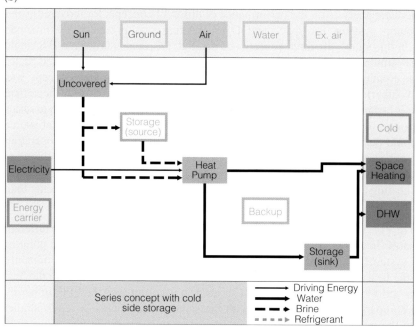

Fig. 7.13 Energy flow charts for a parallel–series system (a) and for a series system (b). Both systems use collectors as the only heat source (single source)

7.2 Series and dual-source concepts

systems turned out to be more cost efficient than using covered collectors despite slightly lower SPF$_{SHP}$ achieved for identical area of the collector field. The control strategy was identical for glazed and unglazed collectors. So, no specific control strategy is used for using unglazed collectors to take advantage of the heat gain with ambient air. The influence of the size of the cold side storage increased with increasing collector areas.

Lerch et al. [18] compared several solar and heat pump systems against reference "heat pump only" systems for the SFH45 building in the climate of Graz. For T44A38, the same systems were simulated again for the climate of Strasbourg [19]. A parallel combination of air source heat pump and 14 m^2 flat-plate solar collectors achieved an SPF$_{SHP+}$ of 3.84 compared with the 2.98 for the "air source heat pump only" system (+0.06/m^2).[15] A single-source P/S system with 30 m^2 of selective unglazed absorbers as the only heat source was analyzed. This system – without cold side storage – performed slightly worse (−0.09 in SPF$_{SHP+}$) than the parallel solar and air source reference with its covered collectors of 14 m^2. Using heat from the same unglazed absorber field for an ice storage, from where it was withdrawn by the heat pump, reduced the SPF$_{SHP+}$ further (−0.12 of SPF$_{SHP+}$). P/S-type systems with 14 m^2 covered collector area and a dual-source heat pump (using collector heat or air) were also investigated. These performed only negligibly better than the parallel solar and air source combination (+0.03 and +0.04 in SPF$_{SHP+}$). It has to be noted, however, that in all series and parallel/series combinations presented the direct collector heat supplied to the combi-storage was slightly lower (around −1%) than that in the parallel solar and air source system, which might be an indicator that the system control could be improved. All solar and air source heat pump combinations were better than the "ground source only" heat pump (SPF$_{SHP+}$ of 3.38), but not as good as a parallel solar and ground source heat pump combination (SPF$_{SHP+}$ of 4.83). The addition of 14 m^2 solar collectors to the ground source heat pump improved SPF$_{SHP+}$ by +1.4.

Mojic et al. [20] compared simulation results from series single-source concepts with selective unglazed collectors for SH and DHW with parallel air source heat pump concepts with 10 m^2 glazed flat-plate collectors. The collector area for the series concepts was chosen based on the criteria of "identical system cost as the reference" and thus 40–80% larger than that for the air source SHP system. The results indicate that the seasonal performance factors for both system approaches (parallel with covered versus series with larger unglazed area) were quite similar in most European climates, with the exceptions of Davos that showed a clear advantage for the series concept and Helsinki that showed a clear advantage for the parallel concept. However, electric backup heating was considerable for these two climates and thus the overall performance was only in the range of SPF$_{SHP}$ = 1.9–2.7. In contrast to the results presented by Citherlet et al. [17], the influence of a cold side storage decreased with increasing collector size. A more costly collector construction with covered collectors that allowed for natural ventilation when operated as a heat source for the heat pump below ambient temperature improved the performance of the overall system only in a few cases.

---

[15] Results discussed here correspond to the results from T44A38 Report C3 [19] Annex G.5 (climate: Strasbourg) rather than to the results presented at the SHC Conference (climate: Graz) [18].

A system with 116 m² unglazed selective absorbers as the only source for a brine source heat pump, integrated in a multifamily house with 927 m² heated floor area located in Geneva, was simulated in TRNSYS and compared with monitoring results by Mermoud et al. [21]. Due to the unusual ratio between SH and DHW (21 and 48 kWh/(m² year)), as well as a hydraulic configuration that does not allow for direct solar preheating of DHW, the simulated seasonal performance factor did not exceed 3.1 (including electricity for the heat pump, and for backup heating when the evaporator temperature drops below −20 °C). With the same total demand, but more usual fractions of 60% SH and 40% DHW, the seasonal performance factor could be increased to 4.6, provided (i) an average meteorological year, without extreme winter temperatures; and (ii) a load regulated heat pump without degradation of the COP. Monitoring results of this system are reported in Section 6.4.

The exclusive use of solar collectors and wastewater heat recovery as a source for a heat pump in P/S-type systems was investigated by Heinz et al. [22]. The simulations show that the SPF_{SHP} of such systems can reach values well above 4, as shown in Table 7.1 for an SFH30 building. The best results were achieved for systems that include an ice storage, which is charged by the collector field and the wastewater heat exchanger (system D). As there is no conventional heat source for the heat pump (air or ground) in the investigated systems, the solar radiation available in winter is crucial for the system performance. This can be seen when the results for the climate Graz are compared with those for Strasbourg, where the solar radiation in winter is significantly lower. In Graz, both a higher SPF_{sys} and lower total electricity consumption are achieved although the space heating demand is significantly higher. A Sankey diagram of system D in Strasbourg for a simulation with T44A38 boundary conditions for space heating is shown in Figure 7.14a.

Table 7.1 Results from Ref. [22] for SFH30 (collector area 30 m²).

Climate/system		Water storage volume (m³)	Ice storage volume (m³)	Recovered wastewater heat (kWh)	Total electricity consumption (kWh)	SPF_{sys}
Graz	A	3	—	486	1953	4.31
	B	3	—	448	2021	4.17
	C	3	—	794	2146	3.93
	D	1.5	1.5	867	1390	5.89
Strasbourg	A	3	—	461	2039	3.53
	B	3	—	477	2048	3.52
	C	3	—	794	2222	3.25
	D	1.5	1.5	825	1378	5.08

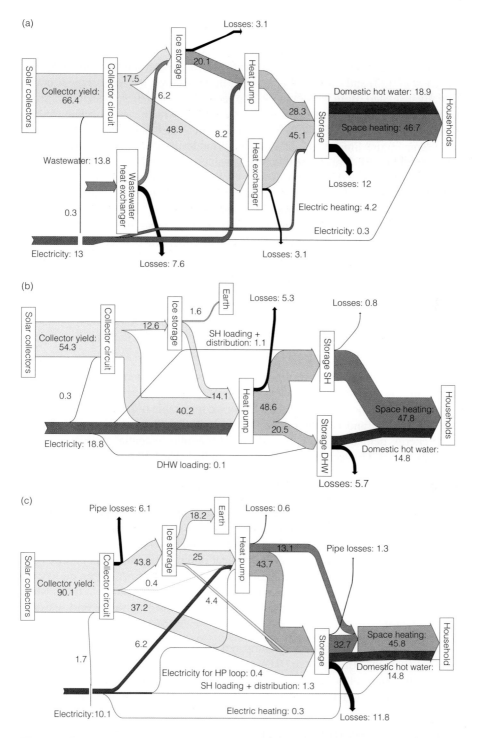

Fig. 7.14 Sankey diagrams of different series concepts with ice storage for an SFH45 building of T44A38 in Strasbourg: (a) solar and waste heat recovery series concept with ice storage by Heinz et al. [22]; (b) solar only series concept with ice storage "small" by Winteler et al. [23]; (c) solar only series concept with ice storage "large" by Carbonell et al. [24]. All values shown are in kWh per m² heated floor surface. Note that a different demand for DHW was assumed for (a) than for (b) and (c)

Single-source system concepts for space heating and DHW preparation for a single-family home with large water/ice storages that are buried in the ground were investigated by Winteler et al. [23] and Carbonell et al. [24]. In these cases, the ice storage is not or only partially insulated. Both ice storage and collector surface in these systems are sized in order to avoid complete freezing of the ice storage, in which case a backup electric heater would be used instead of the heat pump as soon as the source temperature drops below the temperature limit of the heat pump. Winteler et al. [23] found for a series concept in the climate of Strasbourg that SPF_{bSt} values of around 4 are realistic for such systems with ice storage sizes that correspond to about 0.5–2 m^3 per MWh heat demand and (unglazed) absorber areas of about 1 m^2 per MWh heat demand. Carbonell et al. [24] simulated a different type of ice storage and reached SPF_{SHP+} of 5–6 with larger ice storage volumes of 2–4 m^3/MWh and covered collector areas of 2–4 m^2/MWh for parallel and series heat use. For a single-family house with a heat demand of 45 kWh/m^2 and T44A38 boundary conditions, this leads to an ice storage size of ~20 m^3 and a collector area of 30 m^2. Field measurements are currently performed for both system types in order to validate the simulation results. Sankey diagrams of these concepts are presented in Figure 7.14b and c. An open question concerning the long-term performance is, for example, whether the surrounding ground temperatures will change over the years if there is a positive or negative annual net energy flow from the buried storage to the ground, and whether this will change the performance of the system.

The ecological impact of a system with a large ice storage system (SPF of 5.5) was analyzed by means of a life cycle impact assessment (LCIA) and compared with a brine source heat pump with boreholes (SPF of 3.9) for a comparable heat demand.[16] Despite the large ice storage and collector field,[17] the ecological impact in terms of non-renewable energy demand (CED_{NRE}) and global warming potential (GWP) was still dominated by the UCTE electricity mix that was applied. The CED_{NRE} (cradle to grave) of the ice storage system was about 15% lower than that of the borehole system, which was due to the lower electricity demand in operation.

Kurmann et al. [25,26] analyzed a system with parallel and series collector heat use in combination with boreholes and a brine source heat pump. The goal was to find an optimal integration of a 28 m^3 water storage that was used either directly as a heat source at demand temperatures or as a source for the heat pump when the temperatures were lower. The large storage lost also large amounts of heat in the summer. When the system was optimized, the most cost-effective option was to run the heat pump at times when the electricity price was low, despite the fact that this did not lead to the best seasonal performance factor.

Sterling and Collins [27] analyzed and compared three different systems for providing DHW in Ottawa (Canada). The three DHW systems were an exclusively electric system, a solar thermal system (4 m^2 collector area) with electrical backup, and the combined solar and heat pump system that used a large glycol storage (500 l) in the primary circuit

---

[16] See Report C3, Annex H17, of T44A38 [19].
[17] 80 m^3 ice storage, 50 m^2 glazed flat-plate collectors, and 17 m^2 selective unglazed absorbers, compared with 306 m borehole for a heat demand of 36 000 kWh/year.

from where heat can be transferred either with a heat exchanger or by means of a heat pump to the DHW storage. Compared with the electric DHW system, the solar thermal DHW system reduced the electricity demand by 57% and the solar and heat pump combination by 63%.

Citherlet et al. [28] presented a dual-source series DHW system with a brine to water heat pump that used heat from $2\,m^2$ solar collectors (without intermediate storage) or from the space heating loop. For the T44A38 reports, the system was simulated with a second ground source heat pump for space heating. They found that for this system concept (rear side) insulated unglazed selective absorbers perform better than flat-plate, evacuated tube, or uninsulated selective absorbers. The system was also simulated without the solar collectors, using heat from the space heating loop only, which reduced the performance by 6% only. An alternative system where solar heat can also be used in parallel performed better (+14%) than the original system. In this case, evacuated tube collectors were more suitable, and the electricity savings compared with the system without solar collectors were $166\,kWh_{el}$ per m^2 collector area for the climate of Strasbourg.

7.2.4 Multifunctional concepts that include cooling

A SolarCombi+ system that delivers not only space heating and DHW but also space cooling in the climate of Bolzano, Italy, has been simulated by D'Antoni et al. [29]. The system integrated a thermally driven sorption chiller with a reversible compression brine to water heat pump and a relatively large solar thermal field. The electrical heat pump was used for heating and to back up the absorption chiller in cooling operation. Two sources are available for the electrical heat pump: the dry cooler (a heat exchanger with air) and the solar field, and solar heat is used in either parallel or series (P/S dual-source system). The dry cooler is also used to dissipate the energy from the two heat pumps in cooling mode. The calculated SPF_{bSt} (without combi-storage losses) using $28\,m^2$ collector area was 4.5 for Bolzano and 10 for Rome. In this study, the thermal energy used for the calculation of the SPF included cold supply in addition to space heat and DHW.

Chu et al. [30] simulated a dual-source series system for SH, DHW, and space cooling for the climate of Ottawa, Canada, for a Solar Decathlon competition in 2013. A cold propylene–glycol storage was used to collect heat from the solar collectors and provide heat for the heat pump. This cold storage was also charged when cooling was needed, and then cooled by the heat pump if required. A hot storage served for storing heat on the hot side of the heat pump before it was used for space heating or DHW. Alternatively, an outdoor heat exchanger was used for heat rejection in the cooling mode. Heating and cooling were performed with an air-based system for conditioning fresh air from a heat recovery unit as well as recirculated room air. The only heat source of the system – with the exception of waste heat collected from the cooling mode – was the $12\,m^2$ covered collector field. Significant auxiliary heating was needed in December and January. On an annual basis, the system provided 6.0 MWh heat for DHW and space heating, and 7.4 MWh of heat and cold for space cooling and dehumidification (including cooling, dehumidification, and reheating after dehumidification). The latter may be seen as a surprisingly high cooling energy demand for a climate with average maximum daily temperatures of 27 °C for the summer months. If these heat and cold quantities are summed up and divided by the electric energy consumption of 5.8 MWh (27% of this was auxiliary heating), the SPF of the system turns out to be 2.3.

A solar and ground source heat pump concept for DHW, space heating, and cooling has been analyzed by Perers *et al.* [31] for Danish conditions. Based on optimization studies for DHW charging control, it is concluded that weather forecast will be advantageous for the control of this kind of system.

7.3 Special collector designs in series systems

7.3.1 Direct expansion collectors

Direct expansion solar-assisted heat pump water heaters are series systems in which the solar collector is not filled with antifreeze, but with the refrigerant of the heat pump cycle. These systems have been developed since the 1950s [32], and have later been used and introduced into the market, predominantly in warm climates, for domestic hot water preparation. In these climates, SPF for DHW preparation of well above 3 can be achieved as shown by simulations, for example, by Morrison [33]. Within T44A38, Facão and Carvalho reported simulations for such systems in European climates with an SPF of around 2.1 for DHW preparation in Athens.

7.3.2 Photovoltaic–thermal collectors

PVT collectors supply heat and electricity with one component. Accordingly, they can achieve a very high area-specific energetic yield. This aspect can be a crucial factor for projects aiming for high solar fractions with limited roof areas. The combination of PVT collectors in series or dual-source systems is a promising option. Here the heat pump evaporator cools the PVT collector to low temperature levels, which increases the photovoltaic efficiency. However, in periods of high radiation, a heat demand or other heat sink is required to achieve this effect. Therefore, an additional electricity yield can only be achieved by using a cold storage or ground heat exchanger, which allows for the cooling of the PVT in summer.

A series system with unglazed PVT collectors has been measured and simulated for the climate of Kassel, Germany [34]. In this system, the PVT collectors were used for regeneration of the ground in summer. Compared with an uncooled PV array, an additional electrical yield of approximately 4% was simulated on an annual basis and for wind speed that was 50% lower than the free-field meteorological wind speed. The examined systems had freestanding PVT collectors on a flat roof. A collector field size of 1–1.5 m^2 PVT collector per MWh annual heat demand was found to be reasonable. Further analysis and simulations showed that higher additional PV yields of up to +10% can be achieved for PV modules in hotter conditions. Such hotter PV module temperatures can be found for roof-integrated

7.3 Special collector designs in series systems

modules and for low wind speeds. In contrast, for high wind speeds the cooling effect decreases the efficiency gain to 2%.

Dott and Afjei [35] compared different solutions for covering 50 m² of roof area for the T44A38 buildings SFH15–SFH100 with PV (or PVT) and with solar thermal collectors (SC) or absorbers (UC), based on simulations with Polysun:

- PV: 50 m² PV, air–water HP;
- SC: 50 m² solar thermal collectors, air–water HP (P);
- SC + PV: 8 m² solar collectors and 42 m² PV, air–water HP (P);
- PVT: 50 m² PVT collectors as only source of a brine–water HP (P/S);
- UC: 50 m² unglazed polymer absorbers as only source of a brine–water HP (P/S);
- SE: 50 m² selective unglazed absorbers as only source of a brine–water HP (P/S).

Results for the SFH45 building in Strasbourg are shown in Figure 7.15. PV self-consumption was calculated by the integration of simultaneous PV generation and heat pump consumption power over the year, based on hourly time steps.

The amount of electricity purchased from the grid (grid electricity consumption) was lowest for the systems with a high share of solar thermal (covered flat-plate, selective unglazed, or PVT) collectors. The net PV electricity produced was highest for the PVT solution, where the only heat source for the heat pump was the thermal yield of the PVT panels. The amount of electricity purchased from the grid in the coldest winter months (not shown here) was also lowest for the PVT solution, and highest for the air source heat pump, with the other systems at a comparable level in between for the first months

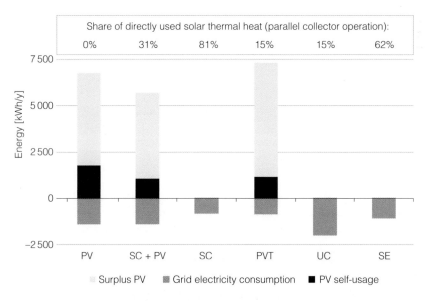

Fig. 7.15 Simulation results for the building SFH45 in Strasbourg, based on Ref. [35]

of the year, and the system with 50 m² solar thermal collectors with a very low electricity demand up to the last few weeks of the year.

7.3.3 Collector designs for using solar heat as well as ambient air

In principle, any unglazed collector can be used also for heat exchange with the ambient air, thus using ambient air as a heat source. An innovative single-source series solar and heat pump concept with a covered flat-plate collector with integrated ventilator has been presented and analyzed in several studies. The energy flow chart for this system is shown in Figure 7.16. The ice storage has a mass of 290 kg water/ice and the combi-storage has a water volume of 1000 l. A simulation study has been performed based on the methodology of EN 12977-2 [36]. A DHW demand of 2945 kWh/year and two space heat loads (A: 9090 kWh/year; B: 6817 kWh/year) were simulated for the climate of Würzburg, Germany. Seasonal performance factors were based on heat delivered after the combi-storage, and included the electricity demand of the heat pump compressor, the air ventilator in the collector, and the source pumps. System seasonal performance factors of 4.9 (25 m² collector area for heat demand A) and 4.8 (20 m² collector area for heat demand B) were obtained.

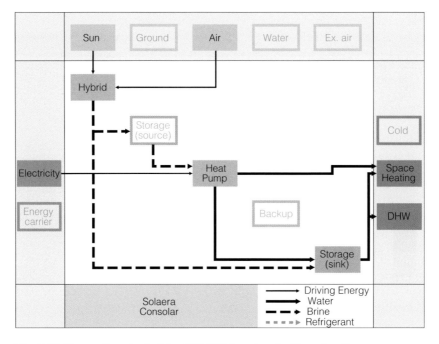

Fig. 7.16 Energy flow chart of the SOLAERA system by Consolar, Germany

7.4 Solar thermal savings versus photovoltaic electricity production

The electricity savings achieved by solar thermal heat use in SHP systems[18] are for DHW systems with low solar fractions[19] slightly larger than the electricity yield from photovoltaic installations in the same Central European climate. For larger solar fractions and for solar combi-systems,[20] this is not the case anymore. However, there are some important differences between systems that are saving electricity by harvesting and storing solar thermal energy on site and systems that compensate electricity consumption of the heat pump by producing electricity on site and feeding it to the electric grid. These differences must be kept in mind when comparing electricity savings by solar thermal and electricity yield of PV and their associated costs:

1. The cost of the solar thermal system usually includes the (full or only additional) cost for storage, whereas for PV no cost is usually calculated for using the grid as a "virtual" storage or for power transfer capacity that is needed. The (additional) cost of this power transfer and virtual storage system is difficult to estimate and depends – last but not least – on the total amount of PV that is feeding to the grid simultaneously. For small fractions of PV generation (compared with total regional or grid-wide electricity generation), the additional cost may be close to zero; for larger fractions, increasing costs have to be expected.
2. Solar thermal yield cannot be used (is lost) if there is not enough demand on site within a short time frame after the harvest, whereas it is usually assumed that the PV yield can be used without limits. The latter may not be true for larger fractions of PV generation (i.e., PV may be stopped from feeding in at times of negative spot market prices).
3. Losses of storage are fully included in the solar thermal savings calculations, whereas losses for grid and – eventually – storages in the grid are difficult to estimate for PV.
4. If PV electricity is fed to the grid and compensated by a feed-in tariff, this usually means that
   a. it gets subsidized (if the tariff is different from the current spot-market price);
   b. the renewable PV electricity is sold to the electricity trader, and so is also the carbon footprint of this electricity.

This means that a PV system whose electricity yield is fed to the grid and reimbursed with a feed-in tariff does not affect the carbon footprint of the local heat pump operation. It affects only the carbon footprint of others who buy this electricity. The carbon footprint of the heat pump operation is defined only by the fraction of PV that is self-consumed plus the electricity mix that is bought additionally from the electricity provider.

For the mentioned reasons, comparing solar thermal and heat pump systems with PV and heat pump combinations is like comparing apples and oranges, unless the PV yield is used locally without being fed to the grid (self-consumption). This can be achieved with batteries or by letting the heat pump operate and charge a thermal energy storage at

---

[18] See Section 7.1.3.
[19] 100–200 kWh per m^2 collector area.
[20] <100 kWh$_{el}$/m^2.

times when there is PV yield. The development of such systems has just started and there is yet little known about the performance as well as about the economics and life cycle cost of these systems.

However, for various reasons comparisons of the currently available technologies of grid-connected PV yield with solar thermal heat pump electricity savings are still interesting and are therefore presented in this chapter.

Dott et al. [37] compared different systems for combined solar DHW preparation with at least 50% fraction for solar thermal or 50% electricity compensation through PV. They found that in terms of ecological aspects as well as economic aspects the two variants are quite similar.

Ochs et al. [38] found for a multistorey passive house in the climate of Innsbruck that small solar thermal systems that cover 20% of the DHW demand are favorable compared with PV from an energetic (efficient use of roof space) point of view. If air source heat pumps are applied, the optimum size of the solar thermal system is higher. The economic evaluation strongly depends on the PV system cost and on the tariff schemes that are assumed for electricity purchase and PV feed-in. The results of this study indicate that highly volatile feed-in tariffs with low prices at times of excessive PV yield and lower prices for feeding in may shift the economic optimum to higher shares of solar thermal collectors than for the feed-in tariffs that are closer to the current situation.

The work of Dott and Afjei [35] that has been referred to in the previous section also provides results for the comparison of PV with solar thermal collectors. The full roof coverage ($50\,m^2$) is at present only interesting from an economical point of view if subsidies or feed-in tariffs are obtained. Large thermal collector fields produce surplus heat in summer that cannot be used, and large PV areas produce much more electricity than is needed for self-consumption and thus depend on a subsidized feed-in tariff or other means of subsidies. From the point of view of an effective use of roof area, the PVT system seems to be the most interesting solution (Figure 7.15).

7.5 Comparison of simulation results with similar boundary conditions

Twelve simulation reports have been published and summarized by different authors where the T44A38 boundary conditions that were defined in the C1 Reports have been applied. This gives the unique opportunity to compare simulation results from different concepts investigated by different authors, since the heat loads, climate, and comfort criteria in these simulations were identical – as shown by the platform independence checks published in Report C4 of T44A38. In this section, only simulation results obtained with T44A38 boundary conditions are compared with each other. The single simulation summary reports are published in Annex G of Report C3, and a summary of these single reports is also included in Sections 7.1–7.4.

7.5 Comparison of simulation results with similar boundary conditions

> For reasons of simplicity, the term "SPF" is used in this chapter for SPF_{SHP+}

In order to include a lot of information in the limited space that is available for figure labels, the following abbreviations are used within this chapter:

hp source – source of the heat pump

- A air source
- G ground source
- S solar collectors

class – classification of the system

- S series (only heat sink for the collector is the heat pump, but the heat pump may use another source too)
- P parallel (solar heat is delivered in parallel to heat from the heat pump)
- R regeneration of ground source
- Ref reference simulation without solar thermal collectors (heat pump only)

coll. type – type of collector

- FL flat-plate collector
- UC uncovered/unglazed absorbers
- SE selective unglazed absorbers

A_{coll} – total collector and absorber area in m^2

V_{st} – total storage volume (source side + sink side) in m^3

ID – identifier for backtracking of the data

The heat load and DHW load for the climate of Strasbourg are summarized in Table 7.2. Design flow and return temperatures of the heating system were 35/30 °C for SFH15 and SFH45, and 55/45 °C for SFH100. For more details, the reader is referred to Report C1.

7.5.1 Results for Strasbourg SFH45

Figure 7.17 shows simulation results for different system concepts serving the heat load of SFH45 in the climate of Strasbourg. Blue bars represent air source heat pump systems, red bars ground source, and orange bars solar collectors as the only source. Green bars indicate systems with a ground source heat pump for space heating in combination with DHW preparation by a second heat pump that uses heat from solar collectors or from the space heating circuit. Reference simulations for heat pump systems without solar thermal collectors are presented with a white filling. It can be seen

Table 7.2 Space heat and DHW load for the different buildings in Strasbourg

Q_{SH} SFH15	2474 kWh
Q_{SH} SFH45	6476 kWh
Q_{SH} SFH100	14031 kWh
Q_{DHW}	2076 kWh

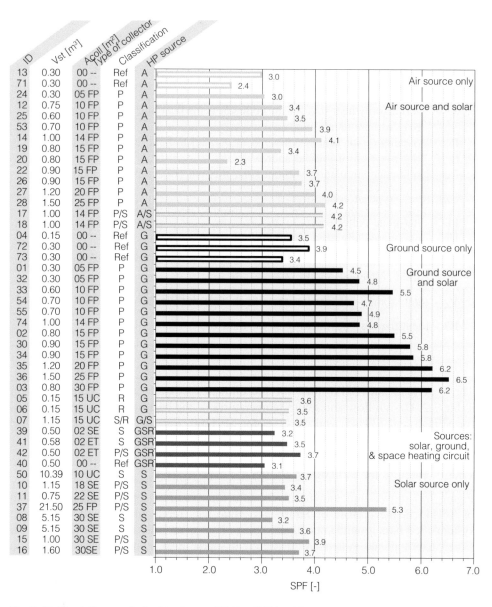

Fig. 7.17 Simulation results for Strasbourg SFH45: SPF for different system concepts with different collector areas and storage volumes. For abbreviations, see the text

that even for those systems there is a range of possible SPF that is achieved. This can be due to different COP of the heat pump used, the sizing and temperature limits of the heat pump, or different assumptions for the exergetic and energetic efficiency of hydraulic integration and storage management.

Based on the data presented in Figure 7.17, the following conclusions can be made.

7.5.1.1 Heat sources

The highest energetic performance (SPF 3.5–6.5) is achieved by ground source systems. Air source systems and "solar only" source systems perform about equally (SPF 3.0–3.8). Outliers can be found: one is a solar and air source heat pump system (20, SPF 2.3) that is very poor because of bad hydraulic integration and control. Another one is a "solar only" system (37, SPF 5.4) that performs considerably better than other "solar only" systems. This can be explained by the large 25 m² collector area[21] in combination with large ice storage of 20 m³ that is buried in the ground. However, the performance of this system is still significantly below the performance obtained by other simulation studies for parallel solar and ground source systems with similar collector size and only 0.8–1.5 m³ of hot storage (35, 36, and 3, SPF 6.2–6.5).

7.5.1.2 System classes

Well-designed solar and *air source* heat pump systems of the *parallel* type with a reasonable collector area of 10–15 m² achieved an SPF of 3.0–4.0. This is about +1.0 in SPF compared with an air source only system (13, SPF 3.0). The *parallel/series combinations with dual-source heat pump that were reported are not able to outperform these systems* (17 and 18, SPF 3.9).

Well-designed solar and *ground source* heat pump systems of the *parallel* type with a collector area of 10–15 m² achieved an SPF of 5.5–5.9. This is about +2.0–2.4 higher compared with the ground source only system (4, SPF 3.5).[22] Two systems with 10 m² flat-plate collector area seem to be underperforming (54 and 55, SPF 3.8–4.1). In these cases, the collector area of 10 m² was used for DHW only, which is a quite larger collector area than usually recommended. *The ground regenerating systems* with unglazed collectors that were reported do not significantly improve the SPF of these single borehole systems. However, in the simulations presented the length of the borehole was decreased from 110[23] to 90 m without glycol storage (and to 70 m with glycol storage) without significant performance decrease.

With the exception of one system, all systems that use *solar heat as the only heat source* used unglazed collectors or selective unglazed collectors. SPF larger than 3.5 was only

---

[21] 20 m² flat plate + 5 m² selective unglazed.
[22] It has to be noted that the ground source only system was not simulated by the same team as the air source only system and that the SPF of the ground source system was surprisingly low compared with the one of the air source system mentioned above.
[23] It has to be noted that the depth of the ground heat exchanger (110 m) was considerable larger than the standard definitions for T44A38 (85 m).

achieved for systems that used $\geq 20\,m^2$ absorber or collector area and $\geq 1\,m^3$ total storage volume. The best energetic performance of these single-source systems was achieved for the combination of flat-plate collectors and selective unglazed absorbers $(20+5\,m^2)$ with ice/water phase change storage of $20\,m^3$ that has already been mentioned previously (SPF 5.4).

7.5.1.3 Dependence on collector size and additional effort

Figure 7.18 shows the dependence of SPF on the collector field size for simulations performed by different authors with task boundary conditions. Results from systems with exclusively unglazed absorbers (selective or not) are shown with circles. A clear dependence can be found for solar and ground source systems and – with lower slope – also for air source systems. This is not the case for "solar only" sourced systems, where the SPF is more or less constant for the range of collector areas simulated. Reasons for this may be that the simulations were performed by different authors who used quite different assumptions for the collector performance (unglazed polymeric tubes, selective unglazed, or glazed flat plate) and collector heat use (S or P/S), for the COP of the heat pump, for the size of cold side storage, and also for the control of these systems.

Figure 7.19 shows SPF versus an effort indicator that should reflect the material effort and cost of the system to some extent and is calculated as the sum of the variables that are weighted with the arbitrary factors shown in Table 7.3. With the weighting factors that were chosen, it can be seen that the energetic reward for additional investment is better for ground source systems than for air source systems, and that for the "solar only" systems the one with covered collectors (SPF 5.4) is about equally rewarding as ground source systems. However, the cost for the large buried ice storage and the cost assumed for ground heat exchangers are crucial for these results and may be dependent on a specific case or location.

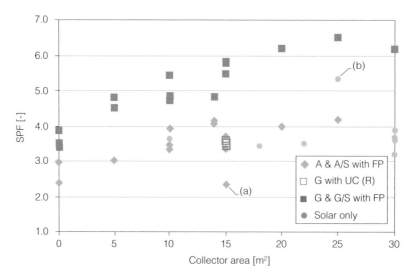

Fig. 7.18 Simulation results for Strasbourg SFH45: SPF versus collector area for different system concepts. Outliers are (a) inappropriate hydraulics and control, and (b) large ice storage and collector field. Abbreviations are explained in section 7.5

7.5 Comparison of simulation results with similar boundary conditions

Table 7.3 Space heat and DHW load for the different buildings in the different climates

Value	Weight A	Per unit
Flat-plate collectors	0.50	/m^2
Selective unglazed absorbers	0.30	/m^2
Unglazed absorbers	0.15	/m^2
Sink storage	2.00	/m^3
Source storage	1.00	/m^3
Air source needed	1.00	/piece
Ground source needed (85 m)	7.00	/GHX
Dual-source heat pump needed[a]	1.00	/piece

[a] Air source systems of the P/S type only.

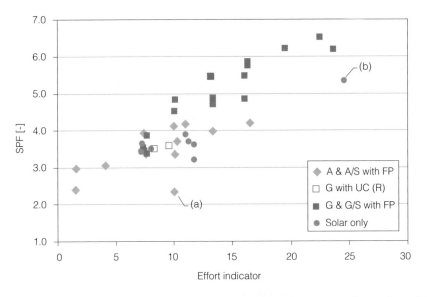

Fig. 7.19 Simulation results for Strasbourg SFH45: SPF versus "effort indicator" for different system concepts. For (a) and (b), see Figure 7.18

7.5.1.4 Electricity consumption

The heat load of the presented simulations for SFH45 in Strasbourg is the same for all systems; therefore, the total electricity consumption is inversely proportional to the SPF values presented above.

The lion's share of the electricity consumption of all simulated solar and heat pump systems can be attributed to the heat pump compressor (Figure 7.20).

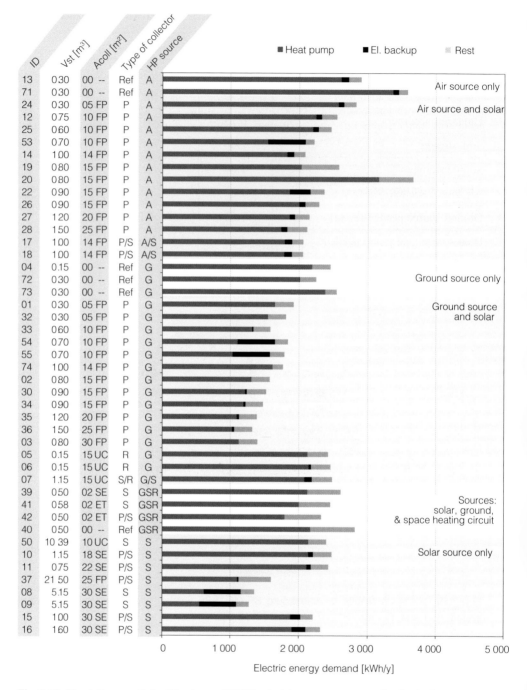

Fig. 7.20 Simulation results for Strasbourg SFH45: electric energy consumption for the heat pump and for the rest of the system. Abbreviations are explained in Section 7.5

7.5 Comparison of simulation results with similar boundary conditions

Depending on the dimensioning of the heat pump and the low-temperature cutoff that was assumed, electric backup heating was responsible for 0–25% of the total electric energy demand for air source systems. Only one author assumed a large share of electric backup for a ground source SHP system, and quite diverging results were obtained for the "solar only" systems. The simulated electric power demand of the controller of these systems ranges from 3 to 30 W. Quite large differences can also be seen in the assumptions for the electricity consumption of pumps for space heating and solar collectors, which can only be explained by quite different assumptions made by different authors for the efficiency of these pumps, the dimensioning and control of these pumps, and/or the pressure drop of the respective fluid circuits.

7.5.2 Results for Strasbourg SFH15 and SFH100

Only few simulations were performed with building loads that are particularly low (SFH15) or particularly high (SFH100). These are presented in Figure 7.21. In general, parallel ground source heat pump systems reached higher SPF values than all other systems.

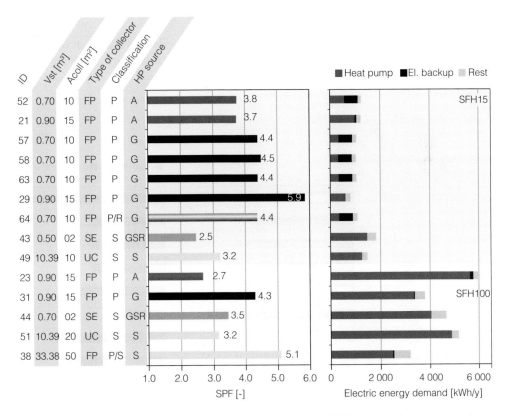

Fig. 7.21 Simulation results for Strasbourg SFH15 and SFH100: SPF and electricity consumption for different system concepts with different collector areas and storage volumes. Abbreviations are explained in Section 7.5

The difference in the overall building energy performance between SFH15 and SFH100 systems is not reflected by the SPF of the system at all, but by the total electricity demand. For SFH15, the lion's share of the electricity demand in many simulations is not the heat pump anymore, but the electric backup heater for cases where the heat pump is only used for space heating and DHW is prepared by solar thermal and electric backup only. In absolute terms, electricity consumers that are additional to the heat pump are not more important for SFH45 than for SFH100, but in relative terms they are responsible for a much higher share of the consumption and thus may have a large influence on the SPF of the system. It has to be noted that the electricity demand for the ventilation with heat recovery that is needed to reach the building standard of SFH15 is not included in the total electricity demand that is presented or in the SPF values shown above.

Obviously, the significance of increasing SPF for SFH100 is entirely different from that for SFH15. For example, increasing SPF_{SHP} from 3.0 to 4.0 saves 1340 kWh_{el}/year for SFH100, but only 380 kWh_{el}/year for SFH15. Thus, additional investments for increasing the SPF are more justifiable for SFH100 than for SFH15.

7.5.3 Results for Davos SFH45

Figure 7.22 shows that energy savings by solar thermal collectors are much higher for Davos than for Strasbourg. For the climate of Davos, the values shown in Figure 7.22 correspond to 170 kWh_{el}/m² for the air source and 130 kWh_{el}/m² for the ground source

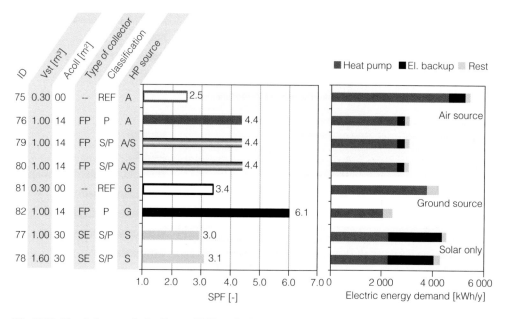

Fig. 7.22 Simulation results for Davos SFH45: SPF and electricity consumption for different system concepts with different collector areas and storage volumes. Abbreviations are explained in Section 7.5

and solar combination. Also for the case of Davos, parallel/series combinations did not outperform the pure parallel concepts. The performance of a "solar only" system was better than the air source heat pump, but not better than the parallel air source and solar combination.

7.6 Conclusions

Parallel concepts with covered collectors that deliver heat to a DHW storage or to a combi-storage are the most used solutions for SHP systems and are simpler in design than series or dual-source systems (see Chapter 1). Compared with a reference heat pump system without solar thermal collectors, simulation results for parallel SHP systems have shown *electricity savings* in the range of 50–200 kWh$_{el}$ per m^2 collector area. For a typical single-family house heat load in a Central European climate, the seasonal performance factor of the overall system increases by about +0.07 to +0.1/m^2 as long as the collector areas are small to moderate. The corresponding area-specific electricity savings are lower the higher the solar fraction. Results are extremely dependent on the climate and – for systems that use solar heat also for space heating – on the availability of solar irradiation during the heating season. The method of *fractional solar consumption can be used with a reasonable accuracy* for a rough estimation of the performance of a known system concept for different climates and heat load situations.

Although parallel solar and heat pump systems seem to be simple solutions compared with the series and dual-source systems, simulations have shown that they cannot be designed by just replacing a fuel-fired boiler of a solar combi-system by a heat pump. Attention has to be paid to the higher mass flow rates of the heat pump compared with fuel boilers of the same heating power, and to the fact that the heat pump's COP is quite sensitive to the temperature level of the heat it is providing. Thus, *exergetic losses on the sink side of the heat pump have to be re-evaluated and minimized for these systems*. This applies particularly for solar combi-storages that must be optimized for low exergetic losses including smart hydraulic integration and a suitable control for charging and discharging, as well as a good stratification of direct storage inlets.

A *comparison of electricity savings by solar thermal on the one hand and PV yield on the other hand* is hardly possible since the functional task provided by a solar thermal system is not directly comparable to that of a PV system. A comparison on an equal footing with solar thermal can be done for PV electricity that is used locally without a detour through the electricity grid. With trends to more self-consumption, systems that store and use PV locally are likely to be produced and used increasingly and thus are a competitor to solar thermal and heat pump combinations. However, only little scientific information has yet been gathered concerning the measured effectiveness of such systems. This will certainly be a topic for research and independent testing in the coming years.

The *potential to increase the overall system performance by parallel/series concepts instead of parallel concepts was shown to be rather limited*. This seems to be due to the fact that a parallel collector operation leads to higher COP values than a series operation unless the irradiance on the collector field is quite low and the collector efficiency would be close to or below zero in parallel operation. Thus, increasing the performance by series operation is only possible at times with very low irradiance on the collector field where little heat can be collected. Consequently, most simulations of series concepts have shown performance results that were only negligibly different from – and sometimes worse than – results for a parallel solar and heat pump concept with essentially the same components and collector field size.

Ground regeneration for a single borehole has not shown significant performance improvements in any of the simulation studies that were evaluated as long as the ground heat exchanger was properly sized. However, several studies have shown that the performance of systems with undersized boreholes can be increased to the level of a properly designed system. This could be particularly interesting for retrofitting of existing boreholes. The combination of PVT collectors with ground regeneration has shown to increase PV yield by 4% in the climate of Kassel, with the potential to increase it by 10% for hotter conditions (e.g., roof-integrated PV and low wind speeds).

Some advantages can be found in *series concepts where the solar collectors are the only heat source of the system*. In these concepts, no ground source or air source heat exchanger is needed and thus

– costs and complexity can be reduced compared with dual-source heat pump systems;
– ground heat exchangers (boreholes and shallow earth collectors) may be avoided, and thus legal restrictions and possible risks or side effects may be reduced;
– air source evaporator units can be avoided, which may reduce noise and improve aesthetics of the system.

Systems with solar collectors as the only heat source tend to be equipped with ice storage. The overall performance that was reported for *systems with unglazed absorbers and ice storage was in the same range as for ground source HP systems without solar collectors*. A performance in the same range as for parallel ground source solar and HP systems was reported for simulations with large covered collector areas combined with large ice storage.

For all SHP concepts, the heat pump compressor is usually by far the largest consumer of electricity (and thus primary energy as well as purchased end energy). If there is an electric backup heater, its electricity consumption may play an important role for systems with undersized components or for boundary conditions that go beyond the capabilities of the heat pump. This can be due to

– "solar only" source SHP systems with too small absorber areas or too small ice storage volume;
– too short ground source heat exchangers;
– extremely low ambient air temperatures in the case of air source heat pumps.

The fraction of electricity that is used for other consumers such as pumps and control is increasing substantially for systems with a low energy demand (e.g., passive houses) and may decrease the SPF of these systems substantially if these consumers are not chosen carefully. It has to be noted, however, that the electric energy demand of an SHP system in a passive house building is always substantially lower than that for a similar or even better system in an SFH45. The importance of electricity consumption of the solar pump increases substantially with the increasing hours of operation in series and ground regenerating systems.

Appendix 7.A Appendix on simulation boundary conditions and platform independence

Finally,

- control issues play an important role and may lead to quite different results for complex systems (e.g., dual-source P/S or P/S/R systems);
- attention has to be paid when simulating new system concepts where components may be operated outside of the usual range (higher evaporation temperatures for heat pumps and temperatures below ambient for solar collectors);
- for most cases, the best practice parallel concepts are a good solution for substantially improving the overall SPF of a heat pump system;
- parallel/series concepts where solar collectors or absorbers are the only heat source are a good way for replacing other sources if this is a relevant criterion for a particular case; control of these systems is not trivial;
- all SHP systems have to be planned and designed carefully, with a minimum of exergetic losses in storage and hydraulics, and taking into account also limits of the components.

Appendix 7.A Appendix on simulation boundary conditions and platform independence

Common boundary conditions have been defined in order to ensure that the results from simulations performed by different researchers on different simulation platforms are comparable. A summary of the most important characteristics of these common boundary conditions is given in this section. Details can be found in Report C1 (parts A and B) of T44A38 [39,40].

Simulations carried out for T44A38 were performed on different platforms such as TRNSYS, IDA ICE, MATLAB/Simulink, and Polysun. The results obtained for the same boundary conditions applied on the different platforms were compared with each other to ensure platform independence.

7.A.1 Climates

The base climates used for the simulation framework are Strasbourg (moderate), Helsinki (cold), and Athens (warm). In addition to these, Davos is used for an extreme mountainous climate and Montreal for an extreme continental climate (see Figure 7.23).

7.A.2 Building load and DHW demand

The building load is based on the simulation of a one-zone building [39] that was derived from previous IEA SHC Tasks 26 and 32. Four different thermal energy performance levels have been used. For easy identification, these buildings have been labeled according to their rounded heating demand in the climate of Strasbourg in $kWh/(m^2\ year)$ as SFH15, SFH45, and SFH100. The annual energetic balance of these buildings for the climate of Strasbourg is shown in Figure 7.24, and a comparison for SFH45 placed in different climates is shown in Figure 7.25. Low-temperature (floor heating) systems have been assumed for SFH15 and SFH45, and a radiator heating

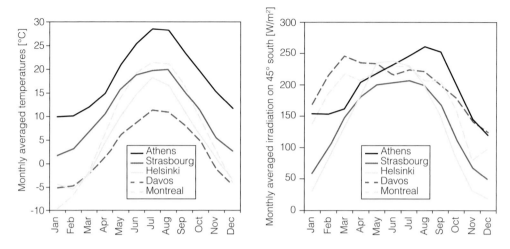

Fig. 7.23 Monthly average values for ambient temperature and solar irradiation on the 45° inclined surface for the climates chosen for T44A38

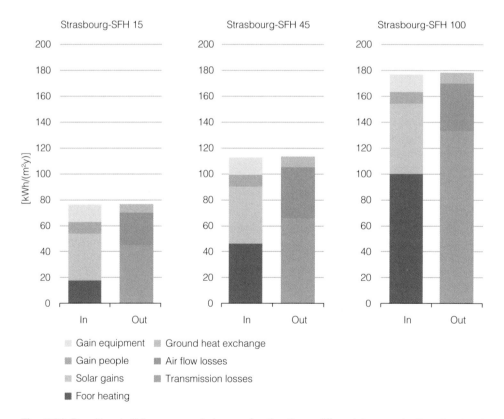

Fig. 7.24 Resulting building energy balances for the three different houses in the climate of Strasbourg

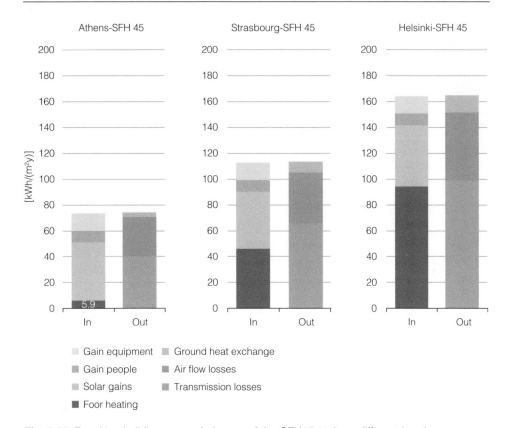

Fig. 7.25 Resulting building energy balances of the SFH45 at three different locations

system with higher temperatures has been assumed for SFH100.[24] Table 7.4 shows the nominal supply and return temperatures of the hydronic heat distribution system, together with the design heating demand, for the three climates of Strasbourg, Helsinki, and Athens.

The supply temperature of the heat distribution system is controlled depending on the 24 h averaged ambient air temperature, and the room temperature set point that is always 20 °C. Thermostatic valves are assumed in all distribution loops for the reference case, reducing the mass flow rate when the room temperature rises, and the heat distribution system is completely off when the 12 h averaged ambient temperature rises above the limit for the heating season.

Seasonal performance factors can be improved considerably in simulation by assuming lower flow and return systems of the space heating system only, without improving the system that delivers the heat. For this reason, a key question for the benchmarking of the central heating unit(s) is the following: What is the temperature

---

[24] The simulation of air-based heat distribution system is excluded here due to a lack of contributions to T44A38 from countries where these systems are predominantly used.

Table 7.4 Building-dependent heating system parameters: design heat load, inlet and outlet temperatures, and heating season limit

Climate	Value	Building		
		SFH15	SFH45	SFH100
All	Ambient temperature limit for heating season (°C)	12	14	15
ST	Space heating power at design conditions (W)	1792	4072	7337
	Supply/return temperature at design conditions (°C)	35/30	35/30	55/45
AT	Space heating power at design conditions (W)	0	1310	3382
	Supply/return temperature at design conditions (°C)	35/30	35/30	55/45
HE	Space heating power at design conditions (W)	3097	6315	10 931
	Supply/return temperature at design conditions (°C)	35/30	40/35	60/50

level required for satisfying the heat demand. Figure 7.26 answers this question by showing for each climate and heat load the cumulative energy delivered to the space heat distribution, versus the maximum temperature of the flow and return line. These curves have been obtained from the hourly averaged heat supply to the space heat distribution system and the hourly averaged values of the inlet and outlet temperatures of the space heat distribution system. For each temperature on the x-axis of the plot, the y-axis value is obtained by integrating the heating power of all hours of the year where the temperature of space heat distribution supply or return was below the respective value. From these plots, it can be derived that, for example, for SFH100 in the climate of Strasbourg the space heating load is 14 MWh/year, and about 6 MWh is delivered with flow temperatures below 40 °C and return temperatures below 25 °C.

> For the comparison of solar and heat pump system performance, these are the most important curves (besides the share and temperature requirements of DHW on the total demand) that two different simulations have to match in order to be comparable!

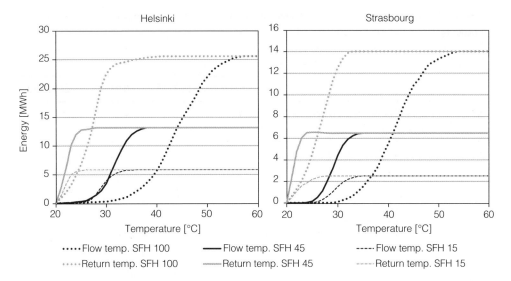

Fig. 7.26 Energy–temperature plots showing cumulative energy demand of the building versus maximum flow and return temperatures for the two locations Strasbourg and Helsinki

The domestic hot water load is based on the EU mandate M/324 tapping cycle M [41]. In order to be suitable for the use in annual simulations, adaptations have been made in order to introduce one bathtub tapping per week, the dependence of hot water demand on the cold water temperature of each location and annual fluctuations. The resulting DHW demand for Strasbourg is 2076 kWh/year, for Helsinki 2398 kWh/year, and for Athens 1648 kWh/year.

In order to assure that the simulated systems are comparable also in terms of comfort provided that manifests in the temperatures of the room and in the DHW temperatures, an "electric energy consumption penalty" is added to the electric consumption of the system whenever the room temperature drops below 19.5 °C or the 2 min averaged DHW temperature drops below 45 °C. Details on this procedure can be found in Report C3 of T44A38 [19].

7.A.3 Other boundary conditions

For the simulation of ground heat exchangers, reference ground properties are defined together with standard heat exchanger design (e.g., number and lengths of probes) for the climates of Strasbourg and Helsinki. The standard simulation uses vertical borehole heat exchangers with double-U pipes, 0.18 m borehole diameter and 0.026 m inner pipe diameter; thermal conductivities of the ground and of the filling material are 2 W/(m K) for the base case.

The number and length of the boreholes are different for each building and each climate as shown in Table 7.5.

Table 7.5 Number and length of borehole heat exchangers for the different locations and buildings

	Strasbourg			Helsinki		
	SFH15	SFH45	SFH100	SFH15	SFH45	SFH100
Maximum heat extraction (kW)	3.5	4.2	7.0	3.5	5.6	10.5
Borehole(s) (m)	49	84	2×90	75	2×95	4×95

7.A.4 Platform validation

Platform validation checks have been performed in order to guarantee similar results for the implementation of boundary conditions on different platforms. Some of the values that were compared in these platform independence checks are the monthly values of direct and diffuse irradiation on the 45° inclined south facing surface, ambient air temperature, domestic hot water energy, and space heating demand. Last but not least, also the energy–temperature plots for the flow and return of the heat distribution systems were compared, which was the most difficult to match since different interpretation of the control affected the return temperatures of the simulation dramatically, and sometimes also the total heat demand. The results of these platform independence checks are included in Report C3 of T44A38 [19].

Acknowledgment

The permission granted by Meteotest (Switzerland) to use the corresponding Meteonorm climate data sets for work within T44A38 is gratefully acknowledged.

References

1. Gabathuler, H.R., Mayer, H., and Afjei, T. (2002) Standardschaltungen für Kleinwärmepumpenanlagen – Teil 1: STASCH-Planungshilfen, im Auftrag des Bundesamtes für Energie.

2. Haller, M.Y., Haberl, R., Mojic, I., and Frank, E. (2013) Hydraulic integration and control of heat pump and combi-storage: same components, big differences. SHC Conference 2013, Freiburg, Germany.

3. Huggenberger, A. (2013) Schichtung in thermischen Speichern – Konstruktive Massnahmen am Einlass zum Erhalt der Schichtung. Bachelor thesis, Institut für Solartechnik SPF, Hochschule für Technik HSR, Rapperswil, Switzerland.

4. Haller, M.Y., Haberl, R., Carbonell, D., Philippen, D., and Frank, E. (2014) SOL-HEAP: Solar and Heat Pump Combisystems, im Auftrag des Bundesamt für Energie BFE, Bern.

5. Carbonell, D., Haller, M.Y., and Frank, E. (2014) Potential benefit of combining heat pumps with solar thermal for heating and domestic hot water preparation. *Energy Procedia*, **57**, 2656–2665.

6. Weiss, W. (2003) *Solar Heating Systems for Houses: A Design Handbook for Solar Combisystems*, Earthscan/James & James.

7. Letz, T., Bales, C., and Perers, B. (2009) A new concept for combisystems characterization: the FSC method. *Solar Energy*, **83** (9), 1540–1549.

8. Haller, M.Y. and Frank, E. (2011) On the potential of using heat from solar thermal collectors for heat pump evaporators. Proceedings of the ISES Solar World Congress 2011, August 28–September 2, Kassel, Germany.

9. Huber, A. and Pahud, D. (1999) Untiefe Geothermie: Woher kommt die Energie. Forschungsprogramm Geothermie, Bundesamt für Energie (BFE), Bern.

10. Bertram, E. (2013) Solar assisted heat pump systems with ground heat exchanger – simulation studies. SHC Conference 2013, Freiburg, Germany.

11. Kjellsson, E., Hellström, G., and Perers, B. (2010) Optimization of systems with the combination of ground-source heat pump and solar collectors in dwellings. *Energy*, **35** (6), 2667–2673.

12. Bertram, E., Glembin, J., Scheuren, J., and Zinterra, G. (2009) Soil regeneration by unglazed solar collectors in heat pump systems. Proceedings of the ISES Solar World Congress 2009, Johannesburg, South Africa.

13. Bertram, E., Glembin, J., Scheuren, J., and Rockendorf, G. (2010) Condensation heat gains on unglazed solar collectors in heat pump systems. Proceedings of the EuroSun 2010 Conference, Graz, Austria.

14. Leibundgut, H. (2012) Sol2ergie: System for ZeroEmission Architecture, Chair of Building Systems/ETH Zürich, Zürich.

15. Wyssen, I., Gasser, L., and Wellig, B. (2013) Effiziente Niederhub-Wärmepumpen und -Klimakälteanlagen. News aus der Wärmepumpen-Forschung – 19. Tagung des BFE-Forschungsprogramms "Wärmepumpen und Kälte", Burgdorf, Switzerland, pp. 22–35.

16. Freeman, T.L., Mitchell, J.W., and Audit, T.E. (1979) Performance of combined solar-heat pump systems. *Solar Energy*, **22** (2), 125–135.

17. Citherlet, S., Bony, J., and Nguyen, B. (2008) SOL-PAC – Analyse des performances du couplage d'une pompa à chaleur avec une instnallation solaire thermique pour la rénovation. Rapport final, Swiss Federal Office of Energy (SFOE).

18. Lerch, W., Heiz, A., and Heimrath, R. (2013) Evaluation of combined solar thermal heat pump systems using dynamic system simulations. SHC Conference 2013, Freiburg, Germany.

19. Haller, M.Y. (2014) System Simulation Reports for the IEA SHC Task 44/HPP Annex 38. A Technical Report of Subtask C – Report C3 – Final.

20. Mojic, I., Haller, M.Y., Thissen, B., and Frank, E. (2013) Heat pump system with uncovered and free ventilated covered collectors in combination with a small ice storage. SHC Conference 2013, Freiburg, Germany.

21. Mermoud, F., Fraga, C., Hollmuller, P., Pampaloni, E., and Lachal, B. (2014) COP5: Source froide solaire pour pompe à chaleur avec un COP annuel de 5 généralisable dans le neuf et la rénovation, Université de Genève, Ofiice fédéral de l'énergie OFEN, Bern.

22. Heinz, A., Lerch, W., Breidler, J., Fink, C., and Wagner, W. (2013) Wärmerückgewinnung aus Abwasser im Niedrigenergie- und Passivhaus: Potenzial und Konzepte in Kombination mit Solarthermie und Wärmepumpe – WRGpot. Report 3/2013, Bundesministerium für Verkehr, Innovation und Technologie, Wien, Austria.

23. Winteler, C., Dott, R., and Afjei, T. (2013) Seasonal performance of a combined solar, heat pump and latent heat storage system. Proceedings of CISBAT 2013 International Conference, EPFL, Lausanne, pp. 1005–1010.

24. Carbonell, D., Haller, M.Y., Philippen, D., and Frank, E. (2014) Simulations of combined solar thermal and heat pump systems for domestic hot water and space heating. *Energy Procedia*, **48**, 524–534.

25. Kurmann, P. (2012) Optimierung der Einbindung eines 28 m3 Wasserspeichers in die Beheizung und WW-Versorgung eines EFH mit WW-Wärmepumpe und Solarkollektoren. M.Sc. thesis, Hochschule für Technik und Architektur Freiburg, Freiburg, Switzerland.

26. Kurmann, P., Mesot, T., and Ursenbacher, T. (2012) Projekt OPTIGEN – Optimierung der Einbindung eines 28m^3 Wasser-Speichers in die Beheizung und die WW-Versorgung eines EFH mit W/W-Wärmepumpe und Solarkollektoren, Bern.

27. Sterling, S.J. and Collins, M.R. (2012) Feasibility analysis of an indirect heat pump assisted solar domestic hot water system. *Applied Energy*, **93**, 11–17.

28. Citherlet, S., Bony, J., Bunea, M., Eicher, S., Hildbrand, C., and Kleijer, A. (2013) Projet AquaPacSol – Couplage d'une pompe à chaleur avec capteurs solaires thermiques pour la production d'eau chaude sanitaire, Office fédéral de l'énergie OFEN.

29. D'Antoni, M., Bettoni, D., Fedrizzi, R., and Sparber, W. (2011) Parametric analysis of a novel SolarCombi+ configuration for commercialization. Proceedings of the 4th International Conference Solar Air-Conditioning, Larnaca, Cyprus.

30. Chu, J., Cruickshank, C.A., Choi, W., and Harrison, S.J. (2013) Modelling of an indirect solar-assisted heat pump system for a high performance residential house. Proceedings of the ASME 2013 7th International Conference on Energy Sustainability (ES2013), ASME, Minneapolis, MN, USA, ES2013-18222.

31. Perers, B., Andersen, E., Furbo, S., Chen, Z., and Tsouvalas, A. (2012) Measurement and modelling of a multifunctional solar plus heatpump system from Nilan.

Experiences from one year of test operation. Proceedings of the EuroSun 2012 Conference.

32. Sporn, P. and Ambrose, E.R. (1955) The heat pump and solar energy. Proceedings of the World Symposium on Applied Solar Energy, The Association for Applied Solar Energy, Stanford Research Institute, Menlo Park, CA, USA, pp. 159–170.

33. Morrison, G.L. (1994) Simulation of packaged solar heat-pump water heaters. *Solar Energy*, **53** (3), 249–257.

34. Bertram, E., Glembin, J., and Rockendorf, G. (2012) Unglazed PVT collectors as additional heat source in heat pump systems with borehole heat exchanger. *Energy Procedia*, **30**, 414–423.

35. Dott, R. and Afjei, T. (2013) System evaluation of combined solar & heat pump systems. Proceedings of CISBAT 2013 International Conference, EPFL, Lausanne, pp. 975–980.

36. Asenbeck, S., Bachmann, S., and Kerskes, H. (2008) Simulationsstudie Solar-Wärmepumpensystem zur Trinkwassererwärmung und Raumheizung. Report Prüfbericht-Nr. 07SIM109/1, Forschungs- und Testzentrum für Solaranlagen, Institut für Thermodynamik und Wärmetechnik, Universität Stuttgart, Stuttgart, Germany.

37. Dott, R., Genkinger, R., Moret, F., and Afjei, T. (2011) Combining heat pumps with solar energy for domestic hot water production. 10th IEA Heat Pump Conference, Tokyo, Japan.

38. Ochs, F., Dermentzis, G., and Feist, W. (2013) Minimization of the residual energy demand of multi-storey passive houses – energetic and economic analysis of solar thermal and PV in combination with a heat pump. International Conference on Solar Heating and Cooling for Buildings and Industry (SHC 2013), September 23–25, Freiburg, Germany.

39. Dott, R., Haller, M., Ruschenburg, J., Ochs, F., and Bony, J. (2013) The Reference Framework for System Simulations of the IEA SHC Task 44/HPP Annex 38 – Report C1 Part B: Buildings and Space Heat Load – Final Revised, FHNW, Muttenz, Switzerland.

40. Haller, M.Y., Dott, R., Ruschenburg, J., Ochs, F., and Bony, J. (2013) The Reference Framework for System Simulations of the IEA SHC Task 44/HPP Annex 38 – Report C1 Part A: General Boundary Conditions – Final Revised. Report C1 Part A, Institut für Solartechnik SPF, Hochschule für Technik HSR, Rapperswil, Switzerland.

41. European Commission (2002) Tren D1 D(2002) M/324 – Mandate to CEN and CENELEC for the Elaboration and Adoption of Measurement Standards for Household Appliances: Water-Heaters, Hot Water Storage Appliances and Water Heating Systems.

8 Economic and market issues

Matteo D'Antoni, Roberto Fedrizzi, and Wolfram Sparber

Summary

This chapter presents the economic analysis of solar and heat pump (SHP) systems in terms of total costs of ownership (investment + running) over a 20-year period. The latter have been adopted to permit a direct comparison with other heating systems (i.e., fossil-fueled systems or district heating systems) and to provide a spendable figure that final users or customers can easily understand.

The outcomes of this chapter testify how SHP systems present clear advantages on environmental and technical levels compared with other systems available on the market: SHP systems are competitive with respect to gas boiler installations, in particular for those applications with a great energy demand such as refurbished or existing buildings or large conditioned areas.

Besides clear advantages on environmental and technical basis, investment costs are a bottleneck for achieving a widespread diffusion of SHP systems on the market. In order to promote a larger diffusion, manufacturers should work on lowering market prices, in the medium to long period. At the same time, European and national incentives should support the diffusion of SHP systems.

Electricity tariffs play an additional role in this scenario. An increase of the electricity tariff would make SHP systems as cost effective as air source heat pumps even for systems with a high investment cost. The increase of the gas price would have a greater impact, in particular for systems with medium to low investment cost.

8.1 Introduction

As seen in Chapters 3 and 7, the performance of SHP systems is influenced by many factors such as location, building loads, and solar collector area. From a theoretical point of view, the limitations of heat pumps and solar thermal collectors can be overcome by their integration into a single energy system. If their combination can reflect a net increase of system seasonal performance factor (SPF) with respect to a traditional heat pump system, the additional ΔSPF has to compensate higher system investment costs. Hence, beneath conducting seasonal energy analysis, an economic analysis is to be tackled to legitimate and to support a wide diffusion of SHP systems.

Since the benefits in terms of energy savings have already been discussed, the intention of this chapter is to focus on economic and environment-related matters. Specifically, the aim is

Solar and Heat Pump Systems for Residential Buildings, First Edition.
Edited by Jean-Christophe Hadorn.
© 2015 Ernst & Sohn GmbH & Co. KG. Published 2015 by Ernst & Sohn GmbH & Co. KG.

- to present a common calculation framework that compares system variants in terms of costs incurred;
- to develop a user-friendly graphical interface (nomograph) for tackling technical, economic, and environmental issues at once and to support designers in decision making.

Hence, the content of this chapter could be a useful support to

- designers, by providing an economic framework based on simple and fair figures that final users/customers can easily understand;
- industry, by indicating target investment costs and minimum performance values to achieve for market competitiveness;
- policy makers, by individuating adequate incentive schemes that could support the diffusion of SHP systems on the market.

8.2 Advantages of SHP systems

The success on the market of HVAC technology is determined mainly by the satisfaction of users. In principle, a good energy system should be cheap and easy to operate and should have limited operation costs, a low maintenance effort, and a long lifespan.

In the last few decades, users have become more sensitive to environmental problems; however, they are not adequately weighted on the energy costs yet, even if clearly motivated. Most of the time, a customer is more attracted by a system with relatively low investment cost rather than a more environment-friendly solution.

International and European norms and standards have redirected the market to environment-friendly systems by imposing an energy labeling to all new HVAC systems and components. This promotes the competition among manufacturers for producing better performing systems at the lowest market cost. Heat pumps and solar thermal collectors are preferential solutions for meeting new restrictive standards on the reduction of CO_2 emissions in new or refurbished residential buildings and, in this sense, SHP systems represent a valuable solution. The large number of ready-to-use SHP systems available on the European market confirms this statement. Kits and turnkey standardized solutions are commercialized in order to reduce the investment costs and the risk of mistakes during installation and to pursue compactness. In order to achieve these results, the standardization of the SHP system layout is fundamental and, thanks to a better design process, this can be reflected in an extended system lifespan.

Integrating solar thermal collectors with heat pump systems shows also positive nonenergetic advantages. For example, adopting solar thermal collectors as an additional heat source reduces the yearly operation time of the device resulting in an increased lifespan once again. Moreover, the noise of air source heat pumps is reduced or eliminated in summer, when customers are more likely to open buildings' windows.

Among all these advantages of SHP systems, energetic and environmental aspects have a leading position. The combination of solar thermal and heat pump technologies contributes to increasing the system SPF and consequently in reducing the final energy (FE) consumption of the system and consequently the CO_2 emissions. The additional

ΔSPF induced by solar collectors is variable and dependent upon the SHP system configuration (series and parallel), the heat source typology, the surface area of the solar field, and the building loads. This relationship has been clearly described in Chapter 7, where a value of ΔSPF per m² of solar field has been given according to parallel and series SHP system layouts. The benefit in terms of FE savings is dependent on the SPF_{ref} of a reference system and on the consequent ΔSPF.

$$FE(\%) = \frac{FE_{ref} - FE_{SHP}}{FE_{ref}} = \left(\frac{1}{SPF_{ref}} - \frac{1}{SPF_{SHP}}\right) \bigg/ \frac{1}{SPF_{ref}}. \tag{8.1}$$

According to this equation and assuming a reference SPF_{ref} value of 2.7 (typical of a pure air source heat pump system [1]) and an SPF_{SHP} value of 4.5, the final energy savings amount to 40%. Since final and consequent primary energy savings are reflected in an equivalent reduction of the CO_2 emissions by using a conversion factor (see Chapter 4), the same value of ΔCO_2 can be achieved. In this sense, SHP systems represent a ready-to-use solution for achieving a renewable energy concept in new or existing residential buildings.

8.3 The economic calculation framework

As declared in the introduction, the aim of this chapter is to support the decision-making process with economic considerations regarding SHP systems for residential applications.

The calculation of the economic benefit of SHP systems has been expressed in terms of energy cost e that the final user has to face within a given period of time. In general, when two systems are compared, the most attractive variant is the one with the lowest value of energy cost e. In order to derive general and replicable outcomes, economic and energy-related quantities have been divided by the building floor area. This calculation procedure has the advantage of being applicable to any system installation that covers space heating and hot water demand.

Any economic analysis focused on energy systems starts from the evaluation of the energy demand. The total energy demand Q_{tot} accounts for space heating (SH) Q_{SH} and domestic hot water (DHW) Q_{DHW} energy demand for residential building applications:

$$Q_{tot} = Q_{SH} + Q_{DHW}. \tag{8.2}$$

Whereas the energy demand for residential applications associated with the DHW load varies typically between 15 and 25 kWh/m² or between 1.2 and 2.4 kWh/day per person [1], the heating demand varies significantly according to building envelope characteristics and climate conditions.

The FE consumption of an electrically driven system can be calculated from the total system seasonal performance factor SPF_{SHP} (see Chapter 4) as follows:

$$FE = \frac{Q_{tot}}{SPF_{SHP}}. \tag{8.3}$$

The FE consumption of a boiler is computed with the overall seasonal efficiency η_{boiler}.

$$\text{FE} = \frac{Q_{tot}}{\eta_{boiler}}. \tag{8.4}$$

The SPF_{SHP} value can be derived from numerical simulations (see Chapter 7) or from the certification of an SHP system. On the contrary, η_{boiler} ranges between 80% (i.e., existing noncondensing gas boilers) and 120% (i.e., gas-driven heat pumps) consumption.

Once the FE consumption has been quantified, the yearly final energy cost C_{fe} can be easily evaluated.

$$C_{fe} = \text{FE} \times u. \tag{8.5}$$

The final energy cost C_{fe} takes only into account the yearly disbursal associated with the energy cost for covering heating and DHW demands. According to the local energy tariff and the energy carrier, the unitary energy price u is meant as the final energy price reported in the energy bill. This unitary price changes with country (see Figure 8.1) and time. In Figure 8.2, the trend of energy price for electricity and gas within EU27 is shown. In case of SHP systems driven exclusively with electricity, an energy price change rate i_e between 4 and 5%/year is found. Furthermore, since the energy tariff might change between day and night, an average value can be derived as follows:

$$u = \frac{\text{FE}_{day} \times u_{day} + \text{FE}_{night} \times u_{night}}{\text{FE}_{day} + \text{FE}_{night}}. \tag{8.6}$$

The amount of money that a family pays for running an SHP system within a certain time frame is defined as the total cost of ownership TCO (€/m^2) calculated according to the net present value (NPV) [2,3] method, which takes into account all costs during the period of analysis:

- initial investment costs I_0;
- consumption-linked payments (final energy costs) C_{fe};
- operation-linked payments (maintenance costs, insurance, and taxes) C_m;
- replacement costs C_r.

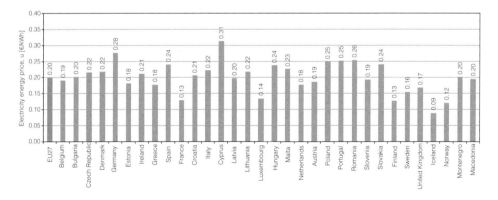

Fig. 8.1 Electricity energy price u in EU27 in the first semester of 2013. (*Source*: Eurostat)

8.3 The economic calculation framework

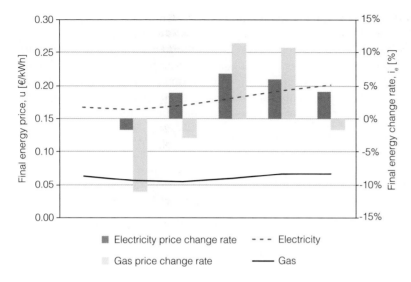

Fig. 8.2 Trend of gas and electricity prices in EU27. (*Source*: Eurostat)

These contributions are then summed according to occurrence time and the associated interest value. In the case of the economic analysis of energy systems, together with the energy price change rate mentioned above, the inflation rate i_{infl} plays a significant role when future disbursals have to be evaluated at the present time. In Figure 8.3, a trend of the inflation costs for the European Community is reported from Eurostat database up to 2012. During the last 10 years, an average value of i_{infl} equal to 2.5% can be noted.

The calculation approach adopted here has assumed that a family can pay the system's investment costs and replacement costs with own budget. Whenever this condition does not occur, these costs are funded through a bank loan. For the sake of simplicity, the loan rate has not been included in the calculations.

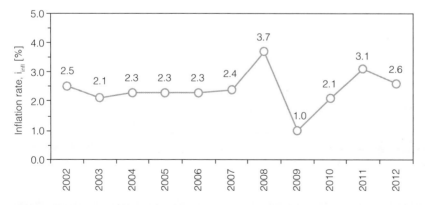

Fig. 8.3 Inflation rate trend in EU27. (*Source*: Eurostat)

In order to compare two investments representing two different energy system variants, a common economic time frame must be defined. This is a key factor in order to compare consistently different system performances under the same economic conditions. Unluckily, there are not many references that guide in assuming a reasonable reference time frame. Typically, active energy systems are assumed to have a lower lifespan (approximately 15–20 years) compared with passive components (30 years) [4]. Therefore, a calculation period N of 20 years is adopted in the following examples.

The advantage of adopting this approach is that the cost effectiveness of a given system is not defined in relative terms with respect to a reference system; on the contrary, it is evaluated in terms of specific energy price that has been paid by a final user. The economic attractiveness is mainly influenced by the amount and the periodicity of disbursals and in this sense initial system investment costs I_0 play a major role.

The SHP system lifespan τ_{SHP} is in general shorter than the calculation period N (Figure 8.5). An estimation of τ_{SHP} is not easy to derive and most of the times it can be based only on personal experience. The system lifespan is not necessarily a function of the lifespan of single components.

When a system completes its lifespan, a replacement occurs. On an economic perspective, this reflects in a series n of replacements, each of them resulting in a replacement cost C_r (see Figure 8.4). Here it is assumed that once the SHP system ends its lifespan, a replacement cost C_r equal to the initial investment I_0 has to be additionally computed. Since replacement costs occur at different times than the initial investment cost, inflation interest i_{infl} has to be considered as follows:

$$\begin{aligned} C_{r,0}^{(1)} &= I_0 \cdot (1 + i_{infl})^{1 \cdot \tau_{SHP}}, & \text{if } 1 \cdot \tau_{SHP} < N, \\ C_{r,0}^{(2)} &= I_0 \cdot (1 + i_{infl})^{2 \cdot \tau_{SHP}}, & \text{if } 2 \cdot \tau_{SHP} < N, \\ &\vdots \\ C_{r,0}^{(n)} &= I_0 \cdot (1 + i_{infl})^{n \cdot \tau_{SHP}}, & \text{if } n \cdot \tau_{SHP} < N. \end{aligned} \quad (8.7)$$

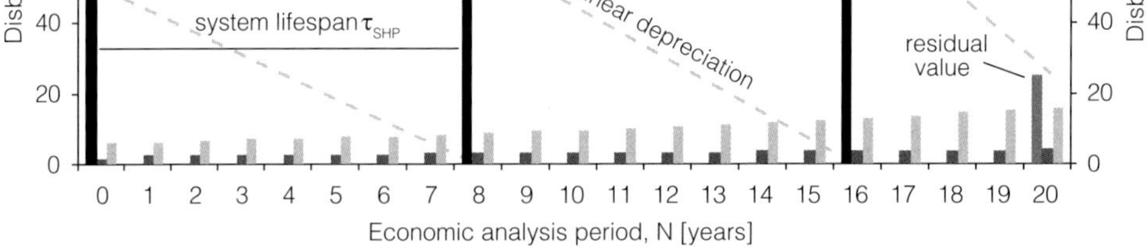

Fig. 8.4 Graphical representation of the periodicity of disbursals and interest-related costs during an economic analysis period

8.3 The economic calculation framework

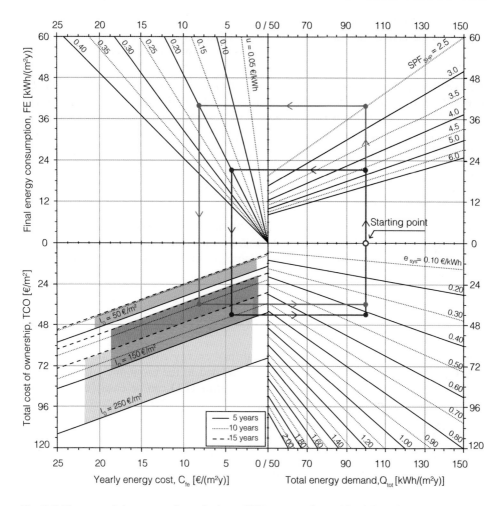

Fig. 8.5 Nomograph for economic analysis on SHP systems for residential applications. Energetic quantities are specified in terms of the living heated area

The total replacement cost $C_{r,0,N}$ is the sum of the single replacement costs that have been faced during the period N.

$$C_{r,0,N} = \sum_{j=1}^{n} C_{r,0}^{(j)}. \tag{8.8}$$

During the lifespan τ_{SHP}, it is assumed that the system has a linear depreciation of the investment cost I_0 or the replacement cost C_r (see Figure 8.4). At the end of the economic analysis period N, a positive residual value RV might occur. The actualized residual value RV_0 of a system can be calculated as follows:

$$RV_0 = \frac{RV}{(1+i_{infl})^N} = \frac{I_0 \cdot (1 - \tau_{SHP}/N)}{(1+i_{infl})^N}. \tag{8.9}$$

Hence, the net total replacement cost $C_{r,N}$ is the difference between the replacement cost $C_{r,0,N}$ and the actualized residual value RV_0 of the system.

$$C_{r,N} = C_{r,0,N} - RV_0. \tag{8.10}$$

The yearly energy-related cost C_{fe} can be calculated as described in Equation 8.5. In order to take into account the total final energy-related disbursal $C_{fe,N}$ that a user will face during the calculation period N, the energy cost change rate i_e has to be taken into account. The total final energy cost $C_{fe,N}$ of a given electrically driven system can be calculated as follows:

$$C_{fe,N} = \sum_{j=1}^{N} C_{fe} \cdot (1 + i_e)^j. \tag{8.11}$$

Since little information from comparable subjects is available, the definition of maintenance cost C_m is also not an easy task. For the sake of simplicity, a benchmark yearly cost is established here as a percentage c_m of the initial system investment cost. This value for SHP systems can range between 1 and 3%/year.

$$C_{m,N} = \sum_{j=1}^{N} C_m \cdot (1 + i_{infl})^j. \tag{8.12}$$

Once the initial investment cost I_0, the total final energy cost $C_{fe,N}$, the maintenance cost $C_{m,N}$, and the net replacement cost $C_{r,N}$ related to the economic analysis period N have been computed, the total cost of ownership TCO can be easily calculated as follows:

$$TCO = I_0 + C_{fe,N} + C_{m,N} + C_{r,N}. \tag{8.13}$$

Therefore, the total system energy price e of an SHP system is simply the ratio between the total cost of ownership TCO and the total energy demand Q_{tot}:

$$e_{SHP} = \frac{TCO}{Q_{tot}}. \tag{8.14}$$

8.4 A nomograph for economic analysis purposes

The nomograph is a two-dimensional graphical calculating interface allowing easy mathematical calculations. Nomographs were extensively used before the spread of computers and calculators. Their adoption today is rare and perceived even odd. Nevertheless, the nomograph is not only a problem-solving aid but also a graphical demonstrator of how many parameters affect a certain problem. In this particular context, a nomograph has been developed with the aim of conducting economic, technical, and environmental analysis at once.

The nomograph consists in a graphical representation of mathematical relationships or laws in which a set of slide rulers providing arithmetic operators such as multiplications

8.4 A nomograph for economic analysis purposes

or divisions are included. Here these rulers represent main SHP system parameters such as the final energy tariff u, the initial investment cost I_0, and the system lifespan τ_{sys}. Before starting to present how the nomograph has been structured and conceived, it is important to declare the field of applicability and the assumptions behind its development.

1. The nomograph should adapt to any future change of
   - system's investment costs;
   - system's performance improvement and technological development related to single components or system interactions;
   - energy cost.
2. It is independent of any reference system definition and geographical locations.
3. It permits to compare energy systems driven by inhomogeneous energy carriers (i.e., electricity and gas), working under the same boundary conditions (location and building loads).
4. It is suited for systems that produce space heating and domestic hot water only, since cooling mode is not considered in T44A38.
5. It applies mainly to residential building cases.

In addition, a set of general assumptions have been implemented:

1. It is assumed that the system's performance will behave constantly (in terms of SPF or η values) during the system lifespan.
2. Inflation i_{infl} and energy price i_e change rate are assumed to be constant during the system lifespan. Two values of 2.5 and 5%/year have been assumed.
3. Maintenance costs are quantified as 2%/year of the initial investment cost.
4. The energy system is assumed to linearly depreciate its value during the system lifespan.
5. The system lifespan for a given system is assumed to be a weighted average among components' lifespans.

In order to use this nomograph, the first step consists in selecting the total building energy demand Q_{tot} on the x-axis of graph 1 (Figure 8.5, upper right). From here, a right-angled line to x-axis should be drawn until the related SPF line is reached. When the SPF_{SHP} value is selected, the final energy consumption is unambiguously calculated (graph 1, Figure 8.5).

A characteristic of the nomograph like this is that the output from a graph becomes the input to the consecutive one. According to this principle, the final energy consumption computed as the output from graph 1 coincides with the input to graph 2 (Figure 8.5, upper left). Here the energy tariff u is considered as the main calculation parameter. This quantity represents the actual electricity energy tariff, not affected by any depreciation or change rate. Similarly to graph 1, a vertical line has to be drawn until the yearly energy cost C_{fe} is returned as output. This quantity represents the cost associated with the yearly energy bills as a function of the parameters (the total heating demand Q_{tot}, the system performance SPF_{SHP}, and the energy tariff u) previously identified.

In graph 3 (Figure 8.5, lower left), the calculation of the total cost of ownership TCO is performed. Two main parameters have to be specified here: the total investment cost I_0 (colored areas) and the system lifespan τ_{SHP} (parallel lines). From a graphical point of view, both are shown in the form of a set of parallel lines where the angular coefficient is a function of the investment cost I_0, and the intercept is dependent on the system lifespan τ_{SHP}. Because of the great amount of parameters in the calculation, the energy change rate i_e, the inflation rate i_{infl}, and the maintenance cost c_m have been assumed as constants. The total cost of ownership TCO accounts these quantities along a simulation period of 20 years.

Differently from the previous graphs, graph 4 (Figure 8.5, lower right) receives two inputs, the total energy demand Q_{tot} (from graph 1, Figure 8.5) and the total cost of ownership TCO (from graph 3, Figure 8.5). The quantities are entered in the horizontal and the vertical axis, respectively, and the intersection point represents the total system energy price e.

The nomograph can be adopted whenever the influence of a calculation parameter (i.e., the system performance SPF_{SHP}) or two different system variants (i.e., air source heat pump only versus a solar-assisted heat pump layout) is compared with respect to the total system energy price e.

The nomograph has not only been conceived for economic analysis on SHP systems. It can be easily changed and adopted for comparing SHP systems and reference fossil-based systems. The main difference consists of the different system performance ratio to adopt in graph 1 (seasonal efficiency performance η_{fossil}) and to the different energy tariff (graph 2, Figure 8.5).

Calculation example

Let us consider, for instance, a single-family house with a total heating demand of $100\,kWh/m^2$. Two SHP systems have been considered for installation:

- *Option 1:* the system SPF_{SHP} is 2.5, the investment cost amounts to $150\,€/m^2$, and the average lifespan is 10 years.
- *Option 2:* the system SPF_{SHP} is 5.0, the investment cost amounts to $250\,€/m^2$, and the average lifespan is 15 years.

What is the total cost of ownership TCO that the dwellings will face in 20 years?

The solution to this problem can be easily derived from the nomograph of Figure 8.5. Once the total heating demand of the building ($100\,kWh/m^2$) is given as input to graph 1 (see "Starting point") and the respective SHP system performances are selected (2.5 and 5.0), the yearly final energy consumption for the variants can be calculated (40 and $20\,kWh/m^2$, respectively). Assuming an average electricity price of $0.20\,€/kWh$ in graph 2, the yearly electricity energy costs associated amount to 8 and $4\,€/m^2$. In graph 3, the system investment costs (colored areas) and lifespans (parallel lines) have to be selected. As output from here the total cost of ownership TCO can be calculated, which accounts for the maintenance cost, replacements cost, and energy inflated costs. In this case, the TCOs amount to 31.6 and 42.3 €/m², respectively. To conclude, the TCO will be divided by the total heating demand so that the heating energy cost the user will face in 20 years can be calculated. The energy cost associated with the first variant is 0.32 €/kWh, whereas the second has reached a higher cost of 0.42 €/kWh. From this simple calculation, it can be concluded that option 1 is more attractive from an economic point of view than option 2.

8.5 Application to real case studies

In this section, the calculation methodology described in Section 8.3 has been applied to real system layout configurations. As an example, a reference fossil-based system and three SHP system variants have been compared in terms of system energy price e. It is pointed out that system investment costs are indicative because of the great variability from country to country. The conclusions of this example do not aim to support or to contest the plausibility of a given system layout, rather provide a calculation comparison among different SHP system variants.

The systems considered are the following:

1. *System variant 1 (VAR1):* condensing gas boiler; investment cost: 50 €/m².
2. *System variant 2 (VAR2):* air source heat pump; investment cost: 70 €/m².
3. *System variant 3 (VAR3):* parallel SHP layout with an air source heat pump + 5 m² flat-plate solar thermal collectors; investment cost: 150 €/m².
4. *System variant 4 (VAR4):* series SHP layout with an air source heat pump + 10 m² unglazed solar thermal collectors; investment cost: 250 €/m².

The m² refers to the space heated area. Each variant has been investigated for different building loads defined as follows:

1. Hot water demand has been assumed equal to 18 kWh/m².
2. Three levels of building loads have been considered: low energy (SFH15), average energy (SFH45), and high energy (SFH100) demand building with a specific load of 15, 45, and 100 kWh/m², respectively.

For the sake of simplicity, the system η_{fossil} value for VAR1 has been fixed to be 1.0. The system performance SPF of an air source heat pump is equal to 2.7 [1]. According to the results of Chapter 7, the SPF_{SHP} values of VAR3 and VAR4 have been assumed to be 3.2 and 3.6, respectively.

The economic analysis has been conducted assuming a calculation time frame N of 20 years, an inflation interest rate i_{infl} of 2.5%/year, an energy price change rate i_e of 5%/year, and maintenance costs equal to 2%/year of the relative initial investment cost I_0. In the case of VAR1 a unitary gas price of 0.10 €/kWh, whereas for SHP systems an

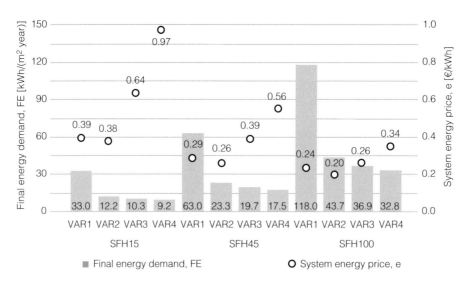

Fig. 8.6 Comparison of different system layout configurations in residential building applications

electrical energy tariff of 0.20 €/kWh has been adopted. Since system standardization in SHP systems is an added value becoming more and more a common practice, a longer lifespan equal to 15 years has been taken for these systems with respect to traditional gas-fired variants (10 years).

In Figure 8.6, the results of the analysis are shown. SHP systems lead to a significant reduction of the final energy consumption of about 63–72% independently of the total heating demand. In absolute terms, the benefit is greater in the case of existing buildings (SFH100) rather than well-insulated new buildings (SFH15), because of the greater absolute final energy savings. In these terms, the benefit brought by SHP system variants (VAR2, VAR3, and VAR4) is quite similar. Since the final energy saving is an inverse function of the system SPF, this result is quite expected because of the narrow variability of SPF values for variants VAR2, VAR3, and VAR4.

Although SHP systems always show a reduction of the final energy consumption, different conclusions should be drawn in terms of system energy prices e. In this regard, the system investment cost I_0 is a decisive parameter in particular for low energy demand buildings. This is due to the fact that system energy price is the ratio of the total cost of ownership TCO and the building energy demand Q_{tot}; therefore, the greater the latter, the lower the system energy price e.

From the previous analysis, it appears that the system investment cost I_0 represents a limit for achieving a satisfying cost effectiveness in an SHP system. At this point, it is interesting to understand which techno-economic conditions allow SHP systems to become as competitive as commercial reference technologies. This exercise evaluates the conditions in which the energy price of a reference system is equal to that of an SHP system.

8.5 Application to real case studies

In order to fill the gap of SHP systems with respect to reference ones on a techno-economic basis, the following measures and scenarios could be considered:

1. an increase of the system SPF_{SHP} value in order to reduce the yearly energy costs;
2. an increase of the energy tariff for the electricity and gas, resulting in increased final energy savings;
3. different scenarios for energy increase costs.

The definition of a reference system is key in this sense, because it strongly influences the conclusions. Because of this, two reference systems have been alternatively considered: a condensing gas boiler (VAR1) and an air source heat pump system (VAR2).

The influence of a variation of the SPF_{SHP} value is shown in Figures 8.7 and 8.8. By selecting the total building energy demand Q_{tot} and the investment cost I_0, the minimum SPF value of an SHP system equating the energy price of a reference system is given on the y-axis. When a gas condensing boiler is considered as a reference system (VAR1) (Figure 8.7), the larger the total energy demand the lower the system performance SPF_{SHP} producing the same cost effectiveness. In Figure 8.8, the same analysis has been conducted using as a reference system an air source heat pump (VAR2).

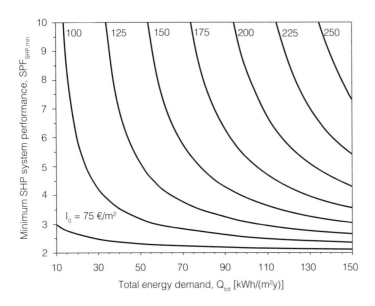

Fig. 8.7 Minimum SPF_{SHP} value of an SHP system for becoming as cost effective as a gas boiler reference system (VAR1). Calculation parameters: electricity tariff: 0.20 €/kWh; gas tariff: 0.10 €/kWh; SHP system lifespan: 15 years; gas boiler lifespan: 10 years; energy increase rate: 5%/year; maintenance cost: 2% of investment cost; inflation costs: 2.5%/year

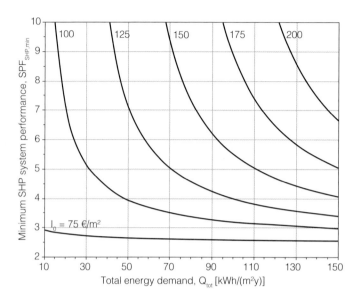

Fig. 8.8 Minimum SPF$_{SHP}$ value of an SHP system for becoming as cost effective as an air source heat pump reference system (VAR2). Calculation parameters: electricity tariff: 0.20 €/kWh; gas tariff: 0.10 €/kWh; SHP system lifespan: 15 years; gas boiler lifespan: 10 years; energy increase rate: 5%/year; maintenance cost: 2% of investment cost; inflation costs: 2.5%/year

Considering, for instance, an SHP system with an investment cost of 100 €/m² and a total energy demand of 60 kWh/m², when this system is compared with the reference system VAR1, the minimum SPF$_{SHP}$ allowing energy cost parity is around 3, whereas in the case of the reference system VAR2 a value of 3.6 is to be achieved. The technical effort in reaching higher values of SPF is increasing exponentially as the energy demand decreases. Hence, an SHP system with an SPF$_{SHP}$ value of 5.0 when compared with a gas boiler system is in general more cost effective for a heating demand Q_{tot} greater than 50 kWh/m² up to an investment cost of about 200 €/m². On the contrary, if the same SHP system is compared with an air source heat pump, the heating demand has to be greater than 70 kWh/m² and the investment cost lower than 175 €/m².

The energy tariff of gas and electricity has a great influence in this analysis. Hence, it is interesting to quantify the energy tariff u that would permit to determine $e_{ref} = e_{SHP}$. In this analysis, the attention is focused on comparing a regular air source heat pump system (VAR2) and an SHP system with a different value of investment costs I_0 (see Figures 8.9 and 8.12).

With respect to an average electricity tariff of 0.20 €/kWh, it is evident that an SHP system with an investment cost of 150 €/m² (Figure 8.9) becomes competitive with an air source heat pump only with SPF$_{SHP}$ values greater than 4.0 and a total energy demand between 90 and 150 kWh/m². If the same analysis is replicated on an SHP

8.5 Application to real case studies

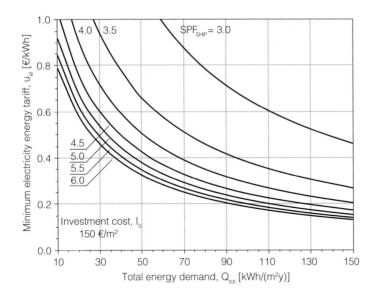

Fig. 8.9 Minimum electricity energy tariff u value for achieving an SHP system (investment cost 150 €/m²) as cost effective as an air source heat pump reference system (VAR2). Calculation parameters: gas tariff: 0.10 €/kWh; SHP system lifespan: 15 years; gas boiler lifespan: 10 years; energy increase rate: 5%/year; maintenance cost: 2% of investment cost; inflation costs: 2.5%/year

system with an investment cost of 250 €/m² (Figure 8.10), the problem has no suitable solution.

Assuming an energy tariff of 0.40 €/kWh for the first case (investment cost of 150 €/m²), cost-effective conditions drop to an energy demand greater than about 40 kWh/m² and SPF_{SHP} value greater than 3.2 approximately. With an investment cost of 250 €/m², optimal solutions are those with an energy demand greater than 100 kWh/m² and SPF_{SHP} greater than 3.5.

Comparing electrically driven systems with gas-driven ones, it is similarly interesting to understand to what extent the gas price can determine the cost effectiveness of SHP systems. With the actual gas price of about 0.10 €/kWh, SHP systems with an investment cost of 150 €/m² (Figure 8.11) are as cost effective as the gas reference system (VAR1) when the energy demand is greater than 65 kWh/m² and for values of SPF_{SHP} greater than 3.0. In the case of an investment cost of 250 €/m², there are no conditions for achieving the same energy cost of the reference system VAR1. In both cases, an increase of the gas price has a positive effect on the SHP system competitiveness.

One of the most difficult parameters to assume during an economic analysis is the energy change rate i_e, because of its intrinsic unpredictable nature; therefore, an

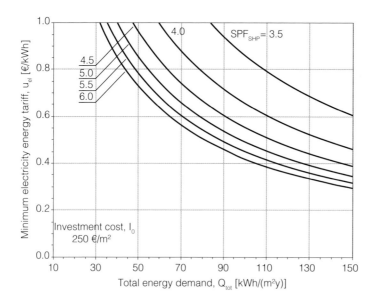

Fig. 8.10 Minimum electricity energy tariff u value for achieving an SHP system (investment cost 250 €/m²) as cost effective as an air source heat pump reference system (VAR2). Calculation parameters: gas tariff: 0.10 €/kWh; SHP system lifespan: 15 years; gas boiler lifespan: 10 years; energy increase rate: 5%/year; maintenance cost: 2% of investment cost; inflation costs: 2.5%/year

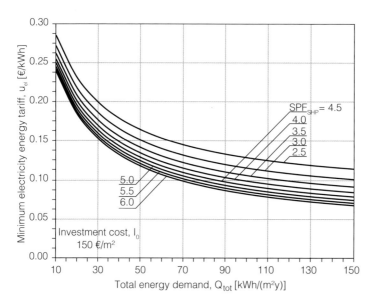

Fig. 8.11 Minimum gas tariff u_{gas} value for achieving an SHP system (investment cost 150 €/m²) as cost effective as a gas boiler reference system (VAR1). Calculation parameters: electricity tariff: 0.20 €/kWh; SHP system lifespan: 15 years; gas boiler lifespan: 10 years; energy increase rate: 5%/year; maintenance cost: 2% of investment cost; inflation costs: 2.5%/year

8.5 Application to real case studies

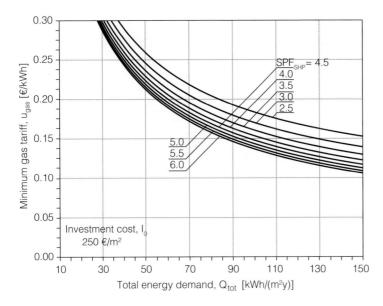

Fig. 8.12 Minimum gas tariff u_{gas} value for achieving an SHP system (investment cost 250 €/m²) as cost effective as a gas boiler reference system (VAR1). Calculation parameters: electricity tariff: 0.20 €/kWh; SHP system lifespan: 15 years; gas boiler lifespan: 10 years; energy increase rate: 5%/year; maintenance cost: 2% of investment cost; inflation costs: 2.5%/year

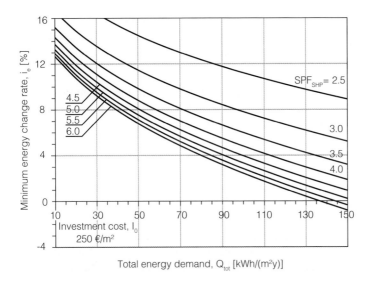

Fig. 8.13 Minimum energy annual change rate i_e for achieving an SHP system (investment cost 150 €/m²) as cost effective as a gas boiler reference system (VAR1). Calculation parameters: electricity tariff: 0.20 €/kWh; gas tariff: 0.10 €/kWh; SHP system lifespan: 15 years; gas boiler lifespan: 10 years; maintenance cost: 2% of investment cost; inflation costs: 2.5%/year

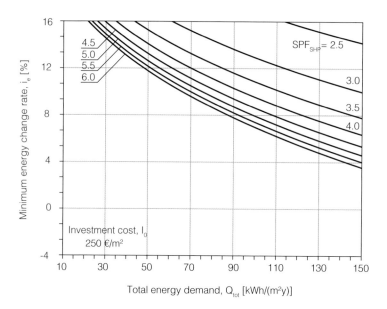

Fig. 8.14 Minimum energy change rate i_e for achieving an SHP system (investment cost 250 €/m²) as cost effective as a gas boiler reference system (VAR1). Calculation parameters: electricity tariff: 0.20 €/kWh; gas tariff: 0.10 €/kWh; SHP system lifespan: 15 years; gas boiler lifespan: 10 years; maintenance cost: 2% of investment cost; inflation costs: 2.5%/year

agreement on this value is always difficult to find. In the previous examples, this value has been fixed to 5%/year according to the actual scenario shown in Figure 8.2. However, it is useful to understand how this scenario would affect the cost effectiveness of SHP systems.

In order to do this, the outcome from this calculation consists of the minimum energy change i_e rate at which an SHP system equals the energy cost per kWh of a reference system. As before, the two reference systems VAR1 and VAR2 and different system investment costs (see Figures 8.13–8.16) have been considered. Also here, the comparison with a gas-fired reference system confirms the great potential of SHP systems. The potential is so great under certain circumstances that they will remain attractive even if the energy change rate would decrease (Figure 8.13). Considering a building energy demand of 70 kWh/m², profitable conditions are reached between i_e values of 4.9–13 and 10.2–17.2% for investment costs of 150 and 250 €/m², respectively. The same calculation with respect to an air source heat pump as a reference system (Figures 8.15 and 8.16) gives 6.7–14.7 and 12.1–19.1%, respectively. This exercise has permitted to understand the huge dependence of the SHP cost effectiveness on future energy scenarios.

It should be pointed out how the dependence on future energy change scenarios is more important for those systems with great final energy consumptions. In order to keep this quantity as low as possible, designers should first reduce the heating energy demand and

8.5 Application to real case studies

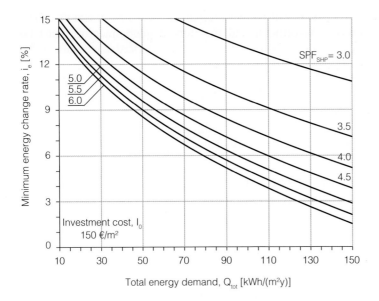

Fig. 8.15 Minimum energy change rate i_e for achieving an SHP system (investment cost 150 €/m²) as cost effective as an air source heat pump reference system (VAR2). Calculation parameters: electricity tariff: 0.20 €/kWh; gas tariff: 0.10 €/kWh; SHP system lifespan: 15 years; gas boiler lifespan: 10 years; maintenance cost: 2% of investment cost; inflation costs: 2.5%/year

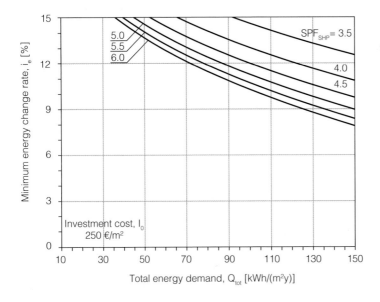

Fig. 8.16 Minimum energy change rate i_e for achieving an SHP system (investment cost 150 €/m²) as cost effective as an air source heat pump reference system (VAR2). Calculation parameters: electricity tariff: 0.20 €/kWh; gas tariff: 0.10 €/kWh; SHP system lifespan: 15 years; gas boiler lifespan: 10 years; maintenance cost: 2% of investment cost; inflation costs: 2.5%/year

then install high-efficiency systems with longer lifespan and low investment cost. The fulfillment of these measures is difficult and therefore a good trade-off must be individuated for achieving not only energy-efficient but also affordable solutions for final users.

References

1. Pezzutto, S. (2012) Analysis of the thermal heating and cooling market in Europe. Proceedings of the 1st International PhD-Day of the Austrian Association for Energy Economics, Vienna, Austria.
2. Verein Deutscher Ingenieure (2000) VDI 2067. Part 1. Economic efficiency of buildings installations. Fundamentals and calculation.
3. Verein Deutscher Ingenieure (1996) VDI 6025: Economy calculation systems for capital goods and plants.
4. European Union (2012) Commission delegated regulation (EU) No. 244/2012. Supplementing Directive 3010/31/EU of the European Parliament and of Council on the energy performance of buildings by establishing a comparative methodology framework for calculating cost-optimal levels of minimum energy performance requirements for buildings and buildings elements.

9 Conclusion and outlook

Jean-Christophe Hadorn and Wolfram Sparber

9.1 Introduction

In order to reduce the CO_2 emissions in the global energy provision next to the electrical energy sector and the transport sector, the heating sector plays a crucial role. In this book, systems have been analyzed that provide heating and domestic hot water for single-family and small multifamily houses exploiting two different renewable energy sources and using electricity as support.

Under the umbrella of the International Energy Agency Solar Heating and Cooling (SHC) Programme and Heat Pump Programme (HPP), a 4-year international collaborative research project has been carried out. The scope of the IEA SHC Task 44/HPP Annex 38 (T44A38) was to analyze the combinations of solar thermal collectors (also including electrical energy production in photovoltaic–thermal (PVT) hybrid collectors) with compression heat pumps in order to satisfy the heating and domestic hot water demand all over the year. The aim was to get an overview of possible combinations and to better understand basic principles and requirements of such hybrid systems, in order to be able to improve them and to achieve high renewable energy fractions.

9.2 Components, systems, performance figures, and laboratory testing

The *single components* used in solar and heat pump (SHP) systems are well known and most of them available in principle on the market since several years. Nevertheless, when applying them in solar and heat pump systems special working conditions might be found that require knowledge of properties that are usually not considered during design and even not tested in standard test methods. An example can be solar thermal collectors applied in a serial SHP system. In this case, the working temperature of the collector will be below the ambient air and its due point for many hours per year, which is not a usual operation condition of a solar collector. Special components might therefore be adapted or developed to achieve an efficient integration in SHP systems.

In order to get an in-depth understanding of the *market-available systems*, in T44A38 an extensive data collection has been carried out. In the years 2010–2012, 80 manufacturers from 11 countries provided information on their systems. A wide variety of different components, hydraulics, and control strategies could be identified to be available. As the market became attractive for a relevant number of companies only in recent years, a consolidation process with regard to systems and hydraulics took place up to now only in a very limited manner.

In order to get clarity on the energy flows when comparing different systems, T44A38 has introduced an efficient way to represent any combination of solar and heat pump systems, with or without storage, in a systematic so-called *energy flow chart diagram*. The diagram simplifies the understanding of a system without losing pertinent information. Figure 9.1 is an example of such a diagram. In Chapter 2, a detailed description

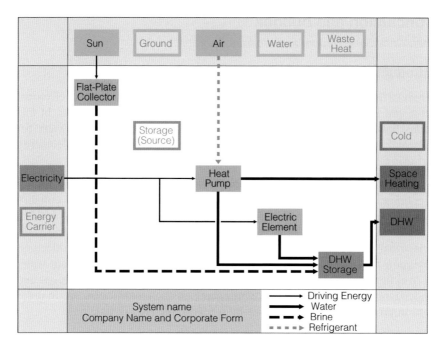

Fig. 9.1 An example of the T44A38 system description tool

of the meaning of the single elements and the energy flows is provided. The chart is available as well as a simple Excel tool from the T44A38 web site.

Furthermore, a *subdivision of the systems* based on the connections between the solar thermal collector field, the heat pump, and the additional heat source has been derived. The categories found have been labeled as parallel (P), series (S), regenerative (R), and complex (C) systems.

As many different systems are available on the market, it is crucial to identify the most relevant energy performance figures and the way to calculate and measure them. The combination of high-quality single components does not necessarily lead to a well-performing system. Next to the electricity consumption of the heat pump itself, the consumption of the auxiliary components (pumps, controllers, displays, fans, motors for valve, etc.) can make a relevant difference.

Based on these experiences within T44A38, the *seasonal performance factor* (SPF) was used as main energy performance figure. Hereby various system boundaries have been defined in order to calculate all relevant performance indicators and the several SPFs that can be defined. The developed method using the energy flow chart of a system can be applied to any kind of systems, not only SHP systems. It allows the comparison of different SHP systems among them as well as a comparison with other heating systems.

Engineers and manufacturers can refer to this work to specify the SPF with common boundaries for the sake of transparent system comparisons.

One way to measure performance figures of SHP systems is *test cycles within dedicated laboratories*. When developing new or improved SHP systems, laboratory testing is regarded as a fundamental and central element. In fact, solar and heat pumps are quite complex systems including dynamical interactions of the main components. In less than a month, relevant information about performances, failures, and default behavior can be obtained from laboratory testing. The delivered information can eventually show clear optimization potential of the tested system.

The laboratory testing process is cost effective since it allows a fast error detection and strongly reduces the risk to position underperforming systems on the market. Such systems can lead to disappointed customers with eventual consequences on reputation and brand confidence. During T44A38, systems showing limited performances that have not been tested prior to market introduction have been observed.

There are different methods under development to test SHP systems in laboratories. Two main approaches are the "single-component testing and system simulation" and the "whole system tests" with data processing, both aiming at estimating performance over a long period of time. The first method can be especially efficient when considering a portfolio of single components and their integration in overall systems. The second method – whole system testing – is characterized by a profound performance characterization of complex systems, by considering system interactions, autonomous controller operation during testing, and close to reality system test sequences.

An internationally recognized quality label for SHP systems does not yet exist. Based on the T44A38 work and experience, it is regarded as an important aspect to develop such a quality label in order to stimulate the development of well-performing systems and to make them recognizable for the customers.

9.3 Monitoring and simulation results and nontechnical aspects

Within T44A38, a special attention has been given to the *measurement and evaluation of installed SHP systems*. In order to do so, it has been agreed on the monitoring strategy, placement of measurement points, and characteristic of measurement devices. Participants of T44A38 have been able to provide monitoring data of periods over 1 or 2 years from 50 different systems in 7 countries. The results show the variety of systems available on the market and also the fact that some systems have still to be considered as prototypes. The variance of results was found to be large with observed system SPFs ranging from 1.5 to 6. As SHP systems are more complex in installation and higher in investment cost than air only source heat pump systems, the performance of SHP systems is expected to be higher too. As a reference, $SPF > 3$ can be used from air source heat pump system monitoring campaigns. In Chapter 6, a detailed description of the monitoring activities and their results are reported. SPFs of several monitored systems have been presented.

The reasons for the variety of results have been analyzed within the T44A38 and are manifold. The complexity of the commissioning of hybrid systems is one important reason. The control strategy is also important. Highly prefabricated kit systems can strongly improve the probability to reach constant high SPF values through extensive testing before market introduction and by strongly reducing the commissioning complexity on site (for instance, by development of "plug-and-play" systems). The hydraulic connections can also be a cause especially regarding the locations of the thermal storage inputs/outputs as shown in Chapter 7.

Regarding the different categories of systems, well-performing systems were found in all four categories (parallel, series, regenerative, and complex) and good integration of all components was shown to be possible. This is the case although parallel systems are the most common and the simplest to operate. Best practice examples are reported in this book.

Within the 4-year period of T44A38, extensive *simulation activities* were carried out. In order to be able to compare different simulation results, a framework for simulating SHP systems in different climates and different loads has been elaborated. The framework is described in detail and can be applied in other projects to compare simulation results of future systems with the results presented in this book. All technical additional task reports are available online for future benchmarking work with other projects. Participants in T44A38 simulated more than 20 different system concepts with the common tools allowing significant comparison activities, under different circumstances.

Chapter 7 gives an overview of simulation results for different SHP system concepts that were applied to a standardized building (SFH45) with a heating demand of 45 kWh/m^2, 150 m^2 living surface, a domestic hot water demand of 6476 kWh, and being located in a Central European climate (Strasbourg). The detailed description of the simulation activities and results is given in Chapter 7. It was found that even for reference systems without solar collectors, there is a range of possible SPFs that can be achieved. This is due to different nominal COP of the heat pump, the sizing of the machine, temperature limits of the heat pump, different assumptions, or even and often different storage management strategies.

It was found that for air source + solar parallel systems the majority of results in terms of SPF are located between 3.0 and 4.2. For ground source + solar parallel systems, SPFs were simulated between 3.5 and 6.2. Very realistically, the reference systems without solar collectors ranked in the first case between 2.4 and 3.0 (air source heat pump) and in the second case between 3.4 and 3.9 (ground source heat pump). Herewith the simulation results are in line with the range of results acquired from monitoring data.

Simulations bring a deeper understanding of the interaction of single components and a better view on critical combinations. Simulations have, for example, shown the great importance of heat storage in a combined system and the necessity of preserving the thermal storage stratification in a tank in order not to diminish the heat pump performance. This aspect turned out to be a key issue for well-performing SHP systems as seen in Chapter 2.

Simulations also showed that the solar contribution benefit in an SHP system can be substantial regarding the reduction of primary energy consumption and greenhouse gas (GHG) emissions.

Finally, based on the simulation results design recommendations have been formulated in this book.

Next to the pure energy performance of SHP systems, evaluations were made with regard to *economic and other nontechnical* aspects. In Chapter 8, nomographs are proposed that allow to assess the economical benefit of single systems and to identify the needed SPF in order to match an equivalent total system energy cost. Furthermore, the monographs allow economic comparisons of SHP with other heating systems.

SHP systems have a higher investment cost than reference systems (air only source heat pump systems or gas boilers have been considered as reference). Through the higher share of freely available renewable energy sources, the energy production is usually less costly. As it has been shown in the economic analysis especially for buildings with a high energy consumption, an economic advantage of SHP systems is possible. Major further factors for the economical liability are the electricity cost, the gas price, and eventual specific penalty costs for nonrenewable energy sources.

Non only the financial comparison but also other aspects can be of interest for final customers and should be considered by systems designers. One of such aspect is given with SHP series systems including unglazed collectors. In such systems, the collector has the double function of large surface air heat exchanger and solar thermal collector. Thanks to this double function, no outdoor unit for the air source heat pump is needed. Therefore, no moving parts have to be installed outside the building, avoiding a source of noise or further possible disturbances.

9.4 Outlook

It can be concluded that solar and heat pump systems will be part of the future energy technologies for space heating and cooling, as well as DHW production, in many countries. If such systems are combined with PV (in parallel or through PVT collectors), they can fulfill the request of a net zero yearly energy balance.

T44A38 has produced the engineering tools to further support the combined SHP technologies (simulation, monitoring, laboratory testing, and performance figure definitions).

It could be observed that high SPFs can be reached in well-integrated systems. Solar heat is a valid source for the heat pump and can bring substantial advantages in terms of economic and energetic considerations.

In the following, single aspects shall be highlighted as they are considered of particular importance.

9.4.1 Energy storage

Within T44A38, it was found that a good integration and operation of the thermal storage is a key issue that can strongly influence the performance of SHP systems. A particular attention to this issue within the development, simulation, testing, and monitoring of systems is regarded as necessary.

In future, when SHP systems are integrated in electricity grids with an increasing share of variable renewable energy sources, the value of the storage could further increase. In fact, SHP systems can store solar energy and thus reduce the need for peak electricity at certain periods of time in a year. Furthermore, they can store thermal energy produced by the heat pump in moments of overproduction of electricity, and therefore contribute to the load management of the electricity grid. A careful policy should promote figures of merit for storing locally, in the house, renewable energy on a hourly and daily basis.

9.4.2 System prefabrication

Several of the systems monitored within T44A38 gave the impression to be custom-made and first of its kind applications. Often the performance results were below expectations. SHP systems are complex hybrid systems with relevant component interaction.

In order to assure a reliable systems performance, to reduce installation time and cost, and to reduce the risk of installation mistakes, compact systems with few outside connections should be developed (plug-and-play approach) and become the standard application within the next few years.

9.4.3 System quality testing

The quality and reliability of systems should be asserted by independent or official bodies. There is, as often for new hybrid technologies, a lack of knowledge on system quality and component interaction and this is a great drawback for customer confidence. It is therefore regarded as crucial that new developed systems are extensively tested in specialized laboratories on a whole system approach.

Internationally accepted quality labels do not exist yet, but are regarded as an important aspect for the future development of SHP systems on the heating market. This book presents some facts and consideration for performance testing methods defining those labels.

9.4.4 Further component development

At the level of components, special designs for SHP systems should be on the to-do list of manufacturers. Topics that the industry could tackle include the following:

- Solar collectors designed for the specific use in SHP.
- Heat pumps with possibly higher temperature on the evaporator side.
- Heat pumps optimized for variable temperature inlet from -10 to $30\,°C$ on the evaporator side.
- Variable conditions for the heat pump (mass flow rates and power, compressor speed) should be easily mastered by compact heat pumps efficient even at all partial loads.
- Solar combistores with better stratification and better control of the way the heat pump charges the store.
- Better understanding of the strategy for the shared heat storage between solar collectors and heat pump.
- The value of local heat or cold storage should be assessed in several grid conditions.

9.4 Outlook

T44A38 has supported the understanding of the integration of solar thermal collectors and heat pump technology and has shown under which conditions the combination of the two technologies can perform efficiently.

The authors of this book hope that through the elaboration of information and results a valuable contribution could be given to the application of solar and heat pump systems on the international heating markets and be considered as a value-added step for further technological development.

Glossary

Abbreviations

AT	Athens
bSt	before storage
BU	backup unit
CCT	concise cycle test
CFD	computational fluid dynamics
COM	cooling operation mode
CTSS	component test system simulation
CU	control unit
DHW	domestic hot water
EHPA	European Heat Pump Association
EN	European Norm
ETC	evacuated tube collector(s)
FPC	flat-plate collector(s)
GHX	ground heat exchanger
HE	Helsinki
HGHX	horizontal ground heat exchanger
HOM	heating operation mode
HP	heat pump
HPP	Heat Pump Programme
HR	heat rejection
HVAC	heating, ventilation, and air conditioning
HX	heat exchanger
IEA	International Energy Agency
ISO	International Organization for Standardization
NPV	net present value
NRE	nonrenewable
OP	operating point
P	parallel system
PCM	phase change material
PVT	photovoltaic–thermal collector
R	regeneration system
S	series system
SC	solar collector(s)
SE	selective unglazed collector(s)
SFH	single-family house
SH	space heating
SHC	Solar Heating and Cooling (Programme)
SHP	solar and heat pump system
SHP+	solar and heat pump system plus energy distribution system
ST	Strasbourg
T44A38	IEA SHC Task 44/Annex 38
TV	thermostatic valve

Solar and Heat Pump Systems for Residential Buildings, First Edition.
Edited by Jean-Christophe Hadorn.
© 2015 Ernst & Sohn GmbH & Co. KG. Published 2015 by Ernst & Sohn GmbH & Co. KG.

UC	uncovered/unglazed
UCTE	Union for the Coordination of the Transmission of Electricity
VDI	Verein Deutscher Ingenieure
VGHX	vertical ground heat exchanger

Symbols

a_1, a_2	loss coefficients for stationary glazed collector model (W/(m² K), W/(m² K²))
A	area (m²)
b_1, b_2, b_u	loss coefficients for stationary unglazed collector model (W/(m² K), J/(m³ K), s/m)
c_{1-6}	parameter coefficients for quasi-dynamic collector model
c_{eff}	area-specific effective heat capacity (J/(m² K))
c_m	maintenance-related cost (%/year)
c_p	specific heat capacity (J/(kg K))
C_{fe}	final energy costs referred to space heated area (€/m²)
C_m	maintenance costs referred to space heated area (€/m²)
C_r	replacement costs referred to space heated area (€/m²)
CED	cumulative energy demand (kWh/kWh)
COP	coefficient of performance
e	total system energy price (€/kWh)
$\dot{E}$	energy flux (W)
EER	energy efficiency ratio
EWI_{sys}	equivalent warming impact of the system (kg CO_2 equiv./kWh)
f_{sav}	fractional savings (%)
f_{sol}	solar fraction (%)
FE	final energy consumption (kWh)
g	gravitation (m/s²)
G	solar irradiance (W/m²)
GWP_{ec}	global warming potential of an energy carrier (kg CO_2 equiv./kWh)
h	specific enthalpy (J/kg)
H	enthalpy (J)
HSPF	heating seasonal performance factor
i	interest (%/year)
I	investment referred to space heated area (€/m²)
$k_{\theta,b}$	IAM for solar beam radiation
$k_{\theta,d}$	IAM for solar diffuse radiation
m	mass (kg)
$\dot{m}$	mass flow rate (kg/s)
N	economic analysis time frame (years)
NTU	number of transfer units
p	pressure (Pa)
p_{el}	specific electric power (W/m²)
P	perimeter (m)
P_{el}	electric power (W)
PE	primary energy (kWh)

Glossary

PEEF	primary energy effort figure
PER	primary energy ratio
$\dot{q}$	specific heat flow rate (W/m^2)
Q	heat (kWh)
$\dot{Q}$	heat flow rate (W)
R	thermal resistance (m^2 K/W)
RV	residual value of the system referred to space heated area (€/m^2)
$\dot{S}$	entropy generation rate (J/(K s))
SCOP	seasonal coefficient of performance
SEER	seasonal energy efficiency ratio
SPF	seasonal performance factor
SSE	squares of relative errors (%)
T	temperature (K)
ΔT	temperature difference (K)
u	final energy tariff (€/kWh)
U	overall heat transfer coefficient (W/(m^2 K))
UA	overall heat transfer coefficient area product (W/K)
W	work (J)

Greek letters

α	angle (°)
α_c	convective heat transfer coefficient (W/(m^2 K))
δ	thermal penetration depth of the ground (m)
ρ	density (kg/m^3)
λ	thermal conductivity (W/(m K))
η	efficiency (%)
η_0	zero-loss collector efficiency
θ	temperature (°C)
Θ	incident angle of beam radiation to the collector (°)
σ	Stefan–Boltzmann constant (W/(m^2 K^4))
τ	time (s)
ω	collector utilization ratio

Subscripts

0	referred to a quantity discounted present quantity
adj	adjacent
amb	ambient air
avg	average
b	beam
brine	antifreeze water mixture
circ	circulation
coll	collector
comp	compressor
cond	condensation
C	cooling, low temperature
d	diffuse

day	daytime
desup	desuperheater
dir	direct charging/discharging
eff	effective
el	electrical
fe	final energy
Fl	flow line
g	global
gen	generation
grd	ground
H	heating, high temperature
in	input
ind	indirect heat use
infl	inflation
IAM	incident angle modifier
lat	latent
lim	limit
loc	location
L	long-wave radiation
k	conductive
min	minimum
night	referred to nighttime
pe	primary energy
prim	primary
PV	photovoltaic
rad	radiative, radiation
ref	reference
ren	renewable
Rd	radiator/floor heating system
Rt	return line
sens	sensible
set	set point
sky	sky vault
std	standard
sys	system
S	shortwave radiation
tot	total
ue	useful energy
use	usable
vent	ventilator
vol	volumetric

Index

A
absolute electricity savings 168
absorbed energy 46
"active" cooling 14
air source heat pumps 31, 139, 143, 147, 149, 179, 221
– systems 179, 191
air source systems 191, 192, 195
air-to-refrigerant heat exchanger 30, 31
air-to-water heat pump 137, 139
– COP performance map 33
– defrosting 136
aluminum 43
ambient heat 25
ambient temperature 25, 26
analytical models 39
ANN. See artificial neural network (ANN)
annual thermal performance 120
antifreeze 184
aquifer thermal energy 42
artificial neural network (ANN) 120
– model 121
autonomous controller operation 127
auxiliary energy consumption, of pumps 109
average annual performance 31
average solar irradiation 64

B
backup electric heating 155
backup unit (BU) 87
BAFA subsidies 20
behavioral model 118
"bidirectionally traded" energy 8
black boxes 75
– model 34
Blumberg single-family house 145
borehole 31
– drilling 37
– effective thermal resistance 38
– heat exchangers 32, 150, 151
 – number and length 204
– heat sources 31, 174

– thermal energy storage 42
Bremer Energie Institute (BEI) 20
brine source heat pump 180, 182
brine/water heat pump evaporator 153
bSt. See system boundaries (bSt)
buffer storage 11
building
– dependent heating system parameters 202
– energy balances
 – houses in climate of Strasbourg 200
– energy demand 220
– loads, levels of 219
– total heating demand of 218
buoyancy-driven natural convection 51

C
capacity-controlled compressors 32
carbon monoxide 46
Carnot efficiency 29, 33, 34
cavern thermal energy storage 42
CCT method. See concise cycle test (CCT) method
CED. See cumulative energy demand (CED)
CED_{NRE} and GWP_{ec} for different energy carriers 90
Central European climate 232
CFD simulations 51, 52
chemical heat pump processes 46
chemical storage reaction 46
chiller models 32
Chinese systems 14
circulation pumps 79
climate 64
– chamber 108
– design 8
– solar and heat pump systems 166
closed loop systems 36
coefficient of performance (COP) 3, 28, 66
– extrapolation of 34
– heat pumps 29, 47, 50, 138, 141, 161, 179, 184, 191, 209, 212
– improvements 49
CO_2 emissions 34, 210, 211, 229

Solar and Heat Pump Systems for Residential Buildings, First Edition.
Edited by Jean-Christophe Hadorn.
© 2015 Ernst & Sohn GmbH & Co. KG. Published 2015 by Ernst & Sohn GmbH & Co. KG.

– savings, fractional 91
cold storage 183
collectors 8, 11
– area/heat demand 168
– design 25
– efficiency 50
– field 49
– heat 47, 49
– models, for solar and heat pump systems (nonexhaustive) 29
– solar (*See* solar collector (SC))
– thermal efficiency 67
– type 14, 17
  – surveyed systems by 17
  – and system concept, cross analysis between 18
– yield 49
COM. *See* cooling operation mode (COM)
combined solar and heat pump systems 47
– parallel vs series collector heat use 47
combined space heating 67
combine solar thermal collectors 122. *See also* collectors
combistore 122. *See also* heat pumps
– combi-storage energy balance 148
combi-systems 106
commercial SHP systems 16
companies, survey for pump systems 13, 14
complex systems 79
component-based simulation program
– TRNSYS 120
component-based testing 106
component performance
– coefficient of performance 70
– seasonal coefficient of performance 70
– solar collector efficiency 71
component testing and whole system testing 106
component-wise testing approach (CTSS) 126
– method 120
– test procedure, extension of toward solar and heat pump systems 120
compressor 28, 32
computational fluid dynamics (CFD) 44
concise cycle test (CCT) method 121

concrete 42
condensation 26, 27, 34, 125
– gains, effect of 176
– heat gains 26
– model errors for 57
condenser 14, 143
consumed electricity 85
convective heat loss coefficient 27
convective heat transfer coefficient 26, 27
"conventional" systems with FPCs 19
conversion coefficient (CC) 68
cooling, as a heat sink 34
cooling effect 25
cooling energy 72, 79, 85
cooling functions 12
– of SHP systems 11
cooling operation mode (COM) 83
– defined 83
COP values. *See* coefficient of performance (COP)
corrosion 43
corrugated pipes 15
cost effectiveness 214
Courant number 45
CTSS. *See* component-wise testing approach (CTSS)
cumulative energy demand (CED) 89

D
daily solar irradiation
– with higher SPF values 125
daily SPF as a function of
– daily solar irradiation and average daily air temperature 125
database 12
data collection 12
Davos SFH45: SPF
– and electricity consumption 196
12-day test procedure 122
defrosting 31
desuperheater 33, 83
– for DHW production 83
devices, for space heating 67
dew point temperature 125
3D finite volume methods 44
DHW. *See* domestic hot water (DHW)

Index

DHW-only SHP systems
- limitations applying for test procedures for 112

direct comparison, of CTSS and WST 109
direct electric defrosting 72
direct energy consumption 87
direct systems 82
"distinguishable" systems 14
documentations 12
domestic hot water (DHW) 67
- buffer storage 155
- circulation system energy consumption
  - for SHP and SHP+system boundaries 81
- consumption 156
- demands 212
- direct solar preheating of 180
- distribution system 81
- heating 177
- high-temperature 143
- load depends 211
- operation mode 85
- with parallel air source heat pump 179
- preparation 14, 15, 79, 141
  - electric backup 176
  - without space heating 19
- production 233
- secondary flow circulation
  - for energy consumption 82
- solar and ground source heat pump concept 184
- and space heating 163
- storages 42, 150
- tank 162
- system
  - electric solar thermal dual-source series 182
- tapping 136
- temperatures 203
  - sensor 51
- zone excessive charging of 123
drivers for, reducing natural stratification 51
DST test procedure, extension of
- toward solar and heat pump systems 120
dual-source heat pump 179, 191

E

Ecoinvent database 90
ecolabeling, of energy using products 66
economic analysis 214
- calculation framework 216
- energy change rate 223
- energy price change 219
- nomograph for 218
- periodicity of disbursals and interest-related costs 214
- SHP systems 218
  - advantages of 211
  - nomograph for 215
- of solar and heat pump (SHP) 209
economic benefit, of SHP systems 211
EER. *See* energy efficiency, ratio (EER)
effective thermal capacitance 26
electrical compressor 126
electrical energy consumption 147
electrical heaters 43
electrical heating element 147
electric backup heating 179, 195
electric energy 123
- consumption penalty 203
electric heating
- backup 176
- elements 13, 138, 145
electric home appliances 67
electricity 2, 9, 25, 28, 126
- demand 32, 146
- energy price 212
- meters 136, 138
- power 28, 120
- tariffs 209, 222
electricity consumption 79, 83, 92, 122, 127, 168
- collector areas and storage volumes 195
- for distribution, assessment of 81
- electrical compressor 126
- of GHX pump 41
- of heat pump 138
electricity savings
- collector areas 168
- for ground source and air source solar and heat pump systems 167

energy
- carriers 90
- consumption 126, 220
  - for circulation pump 82, 87
- conversion system 63
- demand 217, 223
- density 41
- efficient products 68, 104
- and environmental performance 65
- flow (See energy flow)
- performance 120
- quality 69
- related cost 216
- related products 67
- savings 34
- source, for evaporator 47
- tariff 217
  - for electricity/gas 221
  - of gas/electricity 222
- transformation system 75
- transforming components 8
energy balance 28, 44, 66, 79, 83, 85, 151
- equation, for thermal energy storage 44
- of a single PVT collector 25
- for a thermal storage 44
energy efficiency
- calculation, European Commission Regulation 68
- ratio (EER) 66
energy flow 9
- charts
  - diagram 229
  - with parallel air source/ground source 160
  - for parallel/series/regenerative dual-source system 175, 178
- crossing system boundary 79
- representation, of SHP system 93
energy–temperature plots
- cumulative energy demand of building 203
EN 12977 series
- contains test procedures 121
enthalpy 28, 44, 46

- temperature curve 44
ENTSO-E European Network of Transmission System Operators for Electricity 89
environmental energy 8
environmental evaluation
- of SHP systems 87
environmental performance evaluation 69
environmental problems 210
environment-related aspects of system operation 63
equivalent warming impact 91
- of SHP system 89
equivalent warming impact of the system (EWI_{sys}) 72
ETCs. See evacuated tube collectors (ETCs)
EU27
- electricity energy price 212
- gas and electricity prices, trend 213
- inflation rate, trend 213
EU Ecodesign and Energy Labeling Directives 104
European Commission's Directives
- on ecodesign 66
European electricity supply mix 89
European Heat Pump Association (EHPA) 20
European production chains 90
European QAiST project 106
European Solar Thermal Industry Federation (ESTIF) 20
European Union Ecodesign Directive for Energy-Related Products 67
evacuated tube collectors (ETCs) 17
evaporation 34
evaporators 17, 28
EWI_{sys} factor 90
exergetic efficiency 33, 50
exergetic losses
- due to 51
- important sources 50
expansion systems, direct 36
experimental validation, of developed model 126

F

FE. See final energy (FE)
FE consumption, of boiler 212
Federal Office of Economics and Export Control (Bundesamt für Wirtschaft und Ausfuhrkontrolle (BAFA) 20
Federal Statistical Office of Germany (Statistisches Bundesamt) 20
field test 133
– measurements 31
final electrical energy 72
final energy (FE) 88, 89
– consumption, of system 90, 210
flat-plate collectors (FPCs) 17, 24
floor heating systems 15, 161
for DHW charging control 184
fossil-based system 219
fossil fuels 89
FPCs. See flat-plate collectors (FPCs)
fractional electricity savings
– vs. FSC 169, 170
fractional energy savings 67, 74
freezing 26
– water 28
frozen temperature sensors 137
FSC method
– fractional energy savings /performance estimation 169
fuel burners 43
fuel economy 104

G

gas boilers 106
– system 222
gas-driven heat pumps 212
gas-fired reference system 226
gas-fueled boilers 13
generic control strategy 109
geothermal heat exchanger 177
geothermal heat flux 173
g-functions 39
GHX. See ground heat exchanger (GHX)
global solar irradiance 155
global warming 32, 89
– potential (GWP) 89, 182
graphite 46

gravel–water thermal energy storage 42
"gray box" models 118
gray energy 89
greenhouse gas (GHG) emissions 63, 72, 89, 232
grid electricity consumption 185
ground depletion, effects of 174
ground heat exchanger (GHX) 36, 37, 151, 176
– combining with solar collectors 41
– horizontal (See horizontal ground heat exchangers (HGHXs))
– inlet
  – maximum, minimum, and average temperatures 152
– models 38
– performance of 38
– simulation of 203
– vertical (See vertical ground heat exchangers (VGHXs))
ground regenerating systems 177, 191
– simulations of 174
ground source heat pump 31, 145, 173
– systems 191
ground source only heat pump 179
ground source systems 192
groundwater heat pump 31
GSM modems 137
GWP. See global warming, potential (GWP)

H

harmonized WST procedure 117
heat balance
– on monthly and annual basis 152
heat conduction 25
heat consumption 41
heat demand 7, 49, 151
heat distribution 170
heat-driven heat pumps 47
heat exchange 25, 37
heat exchangers 36, 45, 50
– on ambient air 153
heat extraction 25
heat flows 92
– and electricity consumptions 94
– rates of ground heat exchanger 151

– reverse 82
heat for domestic hot water (DHW) 24
heat gains 24, 25
– from rain 26
heat generators 13
heating capacity 67
heating demand 143
– single-family house 218
– to storage losses 141
heating operation mode (HOM) 83
– defined 83
heating rate
– of heat pump 42
heating SPF (HSPF) 67
heating systems 8
– residential buildings 21
heat loads 168, 169
– ground source SHP systems 170
– SFH45 simulations in Strasbourg 166
– solar and heat pump systems 166
heat load SFH45, in Strasbourg 166
heat loss 8, 44
heat meters 136, 139
heat production 41, 50
heat pumps 7, 8, 9, 11, 12, 14, 17, 19, 24, 25, 32, 47, 83, 85, 155, 156
– brine loops of 138
– capacity-controlled 34
– characteristics 15
– circuit and the space heat circuit 149
– and combi-storage 164
– compressor 35
– condensers
  – low operating temperatures 50
– conventional 17
– COP of 174
– crucial for combi-storages 43
– cycle 30
  – refrigerant of 184
– deliver 159
– efficiency 18
  – parameters 50
– electricity demand of 36
– evaporator 125
– ground source 36
– "heat pump only" system 165

– heat pump SPF_{HP}
  – performance factors 149
  – with heat source/heat rejection subsystems 77
– installations 15
– limitations of 209
– machines 2
– models 34
  – nonexhaustive 35
– off 156
– operation 50
– performance, extrapolation of 34
– plus solar thermal system 145
– process 28
– production 156
– programme (HPP) 229
– refrigerant cycle 32
– solar collector, and backup unit 77
– and solar thermal collectors 210
– and solar thermal technologies 65
– systems 7, 85, 133, 209, 231
– technologies 104, 235
  – recent developments 31
– testing and performance assessment
  – standards and guidelines 66
– test procedure 121
– test standard EN 16147:2011 123
heat reservoir 173
heat sink 161
– system 79
heat source 32
– circuit 92
– typology 211
heat storage
– sensible 41
heat storing
– advantage 161
– in water storage 161
heat transfer 38, 44
– coefficients 32
– mechanisms 26
– media 136, 138
– rates 46
– resistance 45
heat traps 43

HGHXs. *See* horizontal ground heat exchangers (HGHXs)
high-quality measurement equipment 127
horizontal ground heat exchangers (HGHXs) 31
– characterization 37
– groups 37
– integrated in basement of a building 37
hot gas defrosting 72
HVAC systems 41, 210
HVAC technology 210
hybrid collectors 17, 87
hybrid technology 1
hydraulic 12
– distribution circuit 155
– scheme 9
hydrocarbons 42
hydrogen 42
hydronic
– heat distribution 15
– heating systems 161
hygienic approaches, modern 15

I
ice storages
– freezing of 182
– for solar and heat pump systems 45
– system, ecological impact 182
IEA Heat Pump Program (HPP) 2
IEA SHC Tasks 169, 199
IEA Solar Heating and Cooling Program (SHC) 2
indirect collector 50
inlet jet mixing 51
inlet temperature 32
in situ monitoring
– installation errors 134
Institute for Heating and Oil Technology (Institut für Wärme und Oeltechnik (IWO) 20
Institut Wohnen und Umwelt (IWU) 20
insulation 37, 43
intercomparable performance, methodology for defining
– within T44A3864

International Energy Agency Solar Heating and Cooling (SHC) Programme 229
investment cost 218
isentropic efficiency 32

K
kinetic energy 51

L
laboratory testing 64
– experience from 120
latent heat exchange 25
latent storage technique 45
life cycle impact assessment (LCIA) 182
long-wave radiation exchange 25
low-temperature heat source 7

M
– macroencapsulated PCMs 46
magnetic valves, consumption of 93
main system boundaries
– for reference SHP system for heating and cooling applications 86
market-available systems 229
market transformation 104
mass flow rate 51
mass transfer coefficients 27
measurement data 125
– of accumulated energy over temperature 124
Mediterranean countries, SHP systems 14, 19
mid-season day 155
minimum electricity energy tariff
– total energy demand 224, 225
minimum energy
– change rate
– annual 225
– total energy demand 227
minimum gas tariff 224
– total energy demand 225
models
– based artificial neural networks 121
– for borehole resistanceg-function 38
– choice and reliable 27

- for covered and unglazed PVT
  collectors 28
- horizontal ground heat exchangers 40
  - with closed loops, to solve 40
- and simulation approach 117
  - with "close to reality" test
    sequences 125
- for vertical ground heat exchangers
  (nonexhaustive)g-functions 39
monitored systems
- characteristics of 139
- measured or correlated SPFSHP 142
monitoring 134
- air/water heat pumps 136
- approach 134
- Blumberg single-family house 145
- data logging systems 137
- Dreieich 149
- electricity meters 138
- heat generators' heating 136
- heat meters 137
- heat pump system with uncovered PVT
  collectors and ground heat
  exchanger 149
- installed heat pumps, nominal COP
  of 141
- meteorological data 139
- model-based evaluation 136
- Satigny 154
- Savièse-Granois 152
- SHP system 136
  - with monitoring equipment 135
- single-family house, in Rapperswil-
  Jona 147
- solar and heat pump performance
  - in situ measurements 139
- standards EN 255 and EN 14511 141
- of T44A38 136
- temperature sensors 139

N

National and international standards and
  guidelines
- heat pump performance 66
natural convection heat transfer
  coefficient 27

net heat gain 44
net PV electricity 185
net total replacement cost 216
nonrenewable (PER_{NRE}) and the EWI_{sys} 90
nonrenewable energy demand
  (CED_{NRE}) 182
nonrenewable energy sources 63, 89
nonrenewable part of energy input
  (PER_{NRE}) 72
nonrenewable resources 72
numerical diffusion 45
numerical models 39, 120, 125
numerical simulation 64

O

one-dimensional storage models
- stratification efficiency 45
operating conditions 63, 122
- climate supply temperature average solar
  irradiation 64

P

parallel collector 50
"parallel-only" concept 18
parallel–series–regenerative system 11
parallel/series (P/S) systems
- with dual-source heat pumps 171
"parameter fitting" test sequences 112
parameterization, of physical models 34
parameters, and their application in
  literature 8
"passive" cooling 14
PER. See primary energy ratio (PER)
performance
- assessment 65
  - of system 64, 127
- calculation 127
- of component/system 69
- evaluation 69
  - of system 125
- factors of 12-day test of system 122
- figures (See performance figures)
- loss, for SHP systems 120
- test procedures, different levels of output
  of 119
performance figures 72

– component 70
– defined 65
– efficiency and 68
– fractional energy savings 74
– renewable heat fraction 74
– solar fraction 72
phase change materials (PCMs) 41
phase changing material (PCM) 153
phase transition 45
phenomenological models 118
– within T44A38 119
photovoltaic–thermal (PVT) collectors 25, 174
physical sorption reactions 46
platform validation, checks 204
plug-and-play approach 234
plug-and-play systems 232
plume entrainment 51
polybutylene 36
polyethylene 36
polymers 43
Polysun simulations 185, 199
primary energy 89
primary energy ratio (PER) 68, 89
– of nonrenewable energy sources 89
primary energy savings
– fractional 91
P/S dual-source systems 177
pumping energy 37
pumping power electricity 174
PV cells 25
PV generation 187
PVT collectors 18, 19, 25, 26, 150, 177
– supply heat and electricity heat demand 184
PV yield 188

Q

qualitative energy flow chart 9
quality assurance tests 104
quasi-steady-state performance map models 32

R

radiator heating system 199
radiators 51

real SHP system
– simplified hydraulic representation 92
reference SHP system 75, 76
refrigerants 14, 32
regenerative approach 11
"regenerative only" product 18
relative humidity 26
renewable energy 21, 88
– equipment 126
renewable heat fraction 74, 156
research projects 12
residential building applications
– system layout configurations 220
reverse cycle defrosting 72
reversible heat pump cycle 29
rock bed thermal storage 42
roller test bench 104
rooftop thermosyphon constructions 14
room heating consumption 142

S

Sankey diagram 28, 166, 180
– for ground source SHP system 166
– heat exchanger, stuties 156
– with ice storage 181
SC. See solar collector (SC)
SCOP. See seasonal coefficient of performance (SCOP)
seasonal coefficient of performance (SCOP) 66
seasonal energy efficiency ratio (SEER) 67
seasonal performance factor (SPF) 67, 77, 126, 146, 149, 186, 209, 211, 230
– boundary for determining 141
– depending on share of domestic hot water measured for solar and heat pumps 144
– for ground source, air source solar and heat pump systems 167
– solar fraction 142
– SPF_{bSt} boundaries
 – on monthly and annual basis 146
 – relative distribution of electricity consumption 147
– SPF_{bSt} values 182
– SPF^*SHP performance values for air source heat pumps 143

– for system boundaries defined within
  T44A38 95
– total electricity consumption 193
SEER. *See* seasonal energy efficiency ratio
  (SEER)
sensible heat exchang 25
sensible heat storage 42
– capacity of water 42
– liquids 42
sensors 108
series collector operation 171
series mode 11
"series-only" concept 16
"series-only" systems 18
series runtime potential 173
SFH45 system
– energy balances 201
– heat load, climate of Strasbourg 189
– heat pump only systems 179
– T44A38 building in Strasbourg 181, 189
SFH100 system
– SPF collector areas and storage
  volumes 195
SH. *See* space heat (SH)
shortwave solar radiation 25
SHP systems. *See* solar and heat pump
  (SHP) systems
simplified climate data 64
simplified hydraulic schemes 10
simulation 37, 109
– boundary conditions (*See* simulation
  boundary conditions)
– and modeling 117
– numerical 38
– programs, dynamic 32
– system (*See* system simulations)
simulation boundary conditions
– platform independence 199
  – building load/DHW demand 203
  – climates 199
  – ground heat exchangers 203
  – platform validation 204
single-source series 171
solar and heat pump (SHP) systems 8, 18,
  19, 166, 169, 212, 216, 233, 234
– applicability 112

– of test data evaluation on 113
– assessment, with basic characteristics
– methods for 65
– boundary 79
– characteristic 136
– companies entering market of 13
– components 234
– concepts DHW, solar heat 162
– for cost effective 221
– economic advantage of 233
– ground source 170
– for heating, and cooling applications
  – main system boundaries for
    reference 80
– lifespan 214
– limitations of test approaches for 114
– measurement and evaluation 231
– net present value (NPV) 212
– parameters energy tariff investment 217
– performance assessment of 65
  – principles for definition of system
    boundaries for 78
– with "series-only" concepts 15
– single components used in 229
– technologies 231
– testing and rating standard for 104
– test procedures 106
– with useful energy distribution
  systems 77
– variants
  – application 234
  – application fossil-based system 219
– without useful energy
  – distribution systems 77
  – storage 77
solar and waste heat recovery series 181
solar-assisted heat pump 184
solar circuit 92
solar collector (SC) 2, 12, 15, 17, 23, 25,
  26, 47, 50, 75, 85, 87, 108, 122, 153, 159,
  168
– efficiency 50, 87
– stagnation 11
– theoretical limit 173
– use of 180
– yield 151

solar combi-storage 147
solar combi-system 52, 183
– with natural gas burner 170
– with space heat distribution control 162
solar DHW preparation 188
solar DHW systems 168
solar energy 11, 17, 19
– usable 169
solar fraction 67
– yield 73
solar heat 32, 42, 45, 46, 47, 173, 176, 191, 233
– from antifreeze fluid 42
– for DHW and space heating 163
– source 33
solar installations 143
solar irradiation 9, 25, 88, 139, 169, 171
– on earth's surface 173
– Strasbourg and collector inlet temperature 172
Solar Keymark certification 17
"solar only" sourced systems 191
solar only source systems 197
solar only systems 192, 195, 197
solar photovoltaics 2
solar radiation 23
– fluctuating supply of 177
solar regenerated systems, robustness of 176
solar source heat pump systems 191
solar storage tanks 161
solar subsystem 9
solar thermal and heat pump systems 159
– *in situ* monitoring 133
solar thermal and storage systems
– without heat pumps 133
solar thermal collectors 17, 21, 23, 24, 26, 52, 66, 67, 122, 185, 186, 209, 210, 235
– design 25
– field 147
  – *subdivision of the systems* 230
– low operating temperatures 50
– PVT collector 151
solar thermal cooling 41
solar thermal heat 164
solar thermal savings

– PV yield 188
– solar thermal and electricity yield 187
– *vs.* photovoltaic electricity production 187
solar thermal systems 15, 21, 106, 119, 137
– design 169
– test procedures 106
solar thermal technology 1, 36
solar thermal yield 187
sorption processes 41
space cooling 15
– functions 14
space heat (SH) 211
– area, building loads 219
– consumption 142
– DHW load 193
– distribution 51, 123, 147, 161
  – pump 147
  – systems 161
– load 42
– climate of Strasbourg 202
– loop 43
– useful energies for 79
SPF. *See* seasonal performance factor (SPF)
stand-alone solar thermal systems 136
standard heat exchanger design 203
standardized testing
– and rating procedures 104
stationary collector efficiency 71
statistical analysis
– market-available solar thermal heat pump systems 11
– methods 12
steady-state collector efficiency 71
storage
– devices 41
– of energy
  – in fuels 42
  – seasonal storage 42
– models, phase change materials 48
– SPF_{bSt}, performance factors of system 149
– stratification 50
– tank 51
Strasbourg SFH45/SFH15
– collector areas and storage volumes 195

- electric energy consumption for heat pump 194
- seasonal performance factor
  - simulation results for 190
  - vs. collector area 192
  - vs. effort indicator 193
stratification efficiency 51
stratified thermal energy storage
- thermal mass for modeling 44
supply temperature 64
surveyed systems
- by concept 18
- correlated by collector type
  - and system concept 18
- by function 15
- by source 16
sustainable growth 104
Swedish single-family home systems 176
Swiss weather conditions 27
switching potential 173
system analysis
- approaches and principles 7
- and categorization 7
- graphical representation 8
system boundaries (bSt) 85
- definition of 75
- for performance evaluation 62
system concepts 15
system energy price
- total cost of ownership (TCO) 220
system functions 14
system investment costs 214, 220
system lifespan 214
system performance figures 71
- seasonal performance factor 71
system prefabrication 234
system quality testing 234
system simulations 27
- parallel solar/heat pump systems 159
  - climates and heat loads 168
  - concepts 163
  - FSC method, fractional energy savings........./performance estimation 169
  - performance 166
- series/dual-source concepts 171

- ground regeneration concepts 174
- multifunctional concepts, include cooling 183
- parallel/series concepts, potential 171
- single source 182
- series systems, special collector designs 184
  - direct expansion collectors 184
  - photovoltaic–thermal collectors 184
  - using solar heat with ambient air 186
- with similar boundary conditions 189
  - collector size/additional (cost) effort, dependence 192
  - Davos SFH45 196
  - electricity consumption 193
  - heat sources 191
  - Strasbourg SFH45 189
  - Strasbourg SFH15/SFH100 195, 196
  - system classes 191
systems, with flow circulation 82
systems with/without additional heating, of distribution pipes 81

T

tank-in-tank solar combistorage 147
tapping cycle 125
tapping test cycle "XL", 122
T44A38 project 141, 229, 231
- boundary conditions 182, 188
- buildings SFH15–SFH100, 181
- companies 12
- market survey 112
- SFH15 with MATLAB/Simulink Carnot Blockset 176
- systems 229
- T44A38, simulation activities 232
- web site 14
TCO. See total cost of ownership (TCO)
techno-economic conditions, SHP systems 220
temperature 26
- dependent density 45
- homogenous 44
- inversion 45
- range for solar installations in dwellings 42

– required by heat demand 47
– sensors 136, 139
– of undisturbed ground 36
test approaches, for SHP systems
– supplying DHW and space heating 114
test cycle
– roller test bench 104
testing boundary 106, 107
– boundary of WST for combi-systems 108
– case for CTSS 107
– characterize performance of SHP systems comparison of approaches 110
– WST methods 107
test laboratories 104, 127
test methods
– for characterization, of SHP combi-systems and DHW-only SHP systems 107
– general requirements for 105
test procedures, for SHP systems 103
– current challenges in developing 106
test sequences
– and applied modeling structure
  – on long-term performance prediction, implications 116
– and determination of annual performance 115
thermal activation, of building elements 42
thermal bridges 43
thermal capacitance 26
thermal characterization 121
thermal collectors 23
– covered 24
– yield 166
thermal conduction 51
– and diffusion 51
thermal convective exchange 24
thermal depletion
– ground source heat pump 173
thermal diffusivity 45
thermal energy storage 41
thermal heat conductivity 27
thermal interference 38
thermal mass 44
thermal performance 67
– of heat pump 125

thermal plume 51
thermal resistance 38
thermal storages 44, 161
thermochemical material
– for reaction 46
thermochemical reactions
– and sorption storage 46
thermostatic valves control 161
tidal energy 88
total cost of ownership (TCO) 216, 218
TRNSYS, IDA, ICE, MATLAB/Simulink and Polysun platforms 199

U
ultrasonic heat meters 138
ultrasonic volume flow meters 137
uncovered/unglazed collectors (UCs) 18
unglazed absorber 24
unglazed collectors, selective 24
useful energy 79, 82, 88
– for DHW preparation 81
U-values 43

V
vacuum tube collectors 24
validation 121
– of capacity-controlled heat pump model 34
– of unglazed collector models 26
ventilators 79
vertical ground heat exchangers (VGHXs) 37
– modeling of 38
– vertically drilled boreholes 37
VGHXs. *See* vertical ground heat exchangers (VGHXs)
visualizations 9
– scheme, introductory example for 9
– solar and heat pump systems 10
volumetric efficiency 32

W
wastewater 16
– heat exchanger 180
– heat recovery, use of 180

water heaters 184
water reservoir, as a direct energy source 36
water storage vessels 43
water tanks, storage models for 47
whole system testing (WST) 108
– for combi-systems 109
  – direct extrapolation of results 116
– procedures for combisystems 108

wind convection coefficient 27
wind energy 88
wind speeds 26, 27
– for reference SHP system for heating/cooling applications 87
wood stoves 13
– for horizontal ground heat exchangers (nonexhaustive) 40
WST. *See* whole system testing (WST)